Historic POWDER HOUSES *of New England*

ARSENALS OF AMERICAN INDEPENDENCE

MATTHEW E. THOMAS

THE History PRESS

Published by The History Press
Charleston, SC 29403
www.historypress.net

First published 2013

Manufactured in the United States

ISBN 978.1.62619.242.3

Library of Congress CIP data applied for.

Notice: The information in this book is true and complete to the best of our knowledge. It is offered without guarantee on the part of the author or The History Press. The author and The History Press disclaim all liability in connection with the use of this book.

Dedicated to Sharon, Rebecca, Daniel and Callie, the epitome of all that makes for a wonderful life. Thank you for being such special people.

CONTENTS

ACKNOWLEDGEMENTS

This author wishes to extend his sincere gratitude to all those who so generously contributed time, photos and information for this book. Without their assistance, this book would not be as in-depth and informative as it is.

First and foremost, I wish to thank my dear wife, Sharon, for her incredible support, patience and encouragement. Without it, this and other previous books would never have been completed. Also, thanks to my daughter Rebecca (Thomas) Curtis, who patiently supported me while gathering historical information during much of her youth. Both she and my wife are the greatest gifts any husband or father could ever ask for.

I also wish to extend a special word of thanks to my good friend and historical associate Glenn Knoblock of Wolfeboro, New Hampshire—a dedicated historical researcher and author of several exceptional New England history books. His helpful advice, recommendations and encouragement of my historical pursuits are genuinely appreciated.

Another word of thanks is extended to Jon DellaPenna of Fremont, New Hampshire, who shares a mutual interest in American history and has accompanied me on historical expeditions in search of powder house photos and memorabilia.

Sincere appreciation to Scott Olsen of Dover, New Hampshire, whose dedicated friendship and words of encouragement urged me to persevere despite numerous unexpected diversions while trying to write this book.

Special thanks to the staff at the New Hampshire Historical Society in Concord for providing valuable assistance. Especially to reference librarian William N. Copeley and to curator and publications director Donna-Belle Garvin, both of whom bring extensive knowledge and professional guidance to all those interested in researching New England history. Also sincere thanks to the New Hampshire State Library in Concord; the Baker Library at Dartmouth College, Hanover, New Hampshire; the Milne Special Collections and Archives at the University of New Hampshire Dimond Library in Durham, New Hampshire; Phillips Exeter Academy Library, Exeter, New Hampshire; Portsmouth, New Hampshire Athenaeum; and to the public libraries in the following New Hampshire towns and cities: Concord, Exeter, Fremont, Manchester, Nashua and Portsmouth. Special thanks and gratitude is extended to Mary Stiner of Fremont, New Hampshire for proofreading the manuscript.

Many thanks to the Massachusetts State Historical Society in Boston; the Boston Harbor Islands National Park; Castle Island Association of South Boston, Massachusetts; and the Massachusetts public libraries located in Haverhill, Newburyport, Salem, Dedham, Beverly, Milton, Pittsfield, Tewksbury and Andover. Also, thanks to the American Antiquarian Society in Worcester, Massachusetts, for being a tremendous historical resource.

Appreciation is extended to the Vermont State Historical Society in Montpelier and Barre; the Connecticut State Library in Hartford; the Providence, Rhode Island Public Library; Maine State Historical Society in Augusta; Maine Department of Conservation; Beth Sweet, director at the Hubbard Free Library in Hallowell, Maine; Connie Manoli-Skocay, assistant librarian, Concord, Massachusetts Public Library; Jonathan Hall of Gilmanton, New Hampshire; Sharon Ann Paradis, Augusta, Maine; Marcy Dawkins, Hampstead, New Hampshire; George Cunningham and Gay Moreau, both of North Attleborough, Massachusetts; Frances R. Killian of the Beverly, Massachusetts Historical Society; Tim Golden of the Foxborough, Massachusetts Public Library; Margaret Y. Perkins of the Holliston, Massachusetts Public Library; Betty J. Brown, town clerk of Marblehead, Massachusetts; Marilyn Erickson, Medfield, Massachusetts Public Library; Paul Hurd, president, Medfield, Massachusetts Historical Society; Eillene Piazza, Milton, Massachusetts Public Library; Jenny O'Neill, director of the Westport, Massachusetts Historical Society; Gregory Lane, Haverhill, Massachusetts Public Library; George Yelle, Norton, Massachusetts; Janet Paine, Norton, Massachusetts Public Library; Denise MacAloney, Westminster, Massachusetts; Cheryl Paul-Bradley, director of the Stevens Library in Ashburnham, Massachusetts; Tom Hardiman of the Portsmouth, New Hampshire Athenaeum; Sandy Howe, Cornish, Maine; Kathryn Gallerani and Jack Burrey, Kingston, Massachusetts; Faith Poulin, Rockland, Maine; John Jordan, Manchester, New Hampshire; Frank J. Gagliardi, Millbury, Massachusetts; Jeanne Campbell, York, Maine; Jonathan Metcalf, Edgecomb, Maine; and Iva Arpin, Voluntown, Connecticut.

INTRODUCTION

What is a powder house? Unfortunately, most people have no idea what these forgotten relics of our early American past are. They are mostly round or rectangular brick, granite or wood structures that housed a community's supply of gunpowder, firearms and other ammunition and military supplies. Some of these old powder houses, sometimes also known as powder magazines and gun houses, date back to the 1600s, with the vast majority of them being built in the first quarter of the 1800s. The oldest known powder houses were built in Massachusetts at Salem in 1637, Dedham in 1653, Somerville in 1703–04 and Boston in 1706.

Few people know that these old landmarks were the much-desired object of British military goals at the outbreak of the American Revolutionary War. After all, it was to the powder houses in Somerville, Massachusetts, on September 1, 1774; New Castle, New Hampshire, on December 14 and 15, 1774; and Concord, Massachusetts, on April 19, 1775, that the British were secretly marching (or sailing) in an attempt to capture any and all barrels of gunpowder, firearms and ammunition stored within these structures. British intent was to prevent these valuable military provisions from falling into the hands of American Patriots who were starting to rebel against King George III of Great Britain. As a result, when the British army began marching on Concord and Lexington, Massachusetts, on April 18 and 19, 1775, to confiscate the powder and arms stored in the Concord Powder House, that was the last straw for the Americans, and the War for American Independence against Great Britain (England) began in earnest.

This book attempts to comprehensively document the construction, cost, history, demise, stories and roles these 201 miscellaneous New England powder houses, powder magazines and gun houses played within their communities, states and nation through the centuries as best as can be researched and recorded. In addition to the purpose and roles powder houses played in New England history, learn about the duels, explosions, fatalities, murders, hangings, accidents and celebrations that took place in or around them.

Some powder houses have detailed information, while others have scant documentation culled from various town histories, newspapers, state histories, state laws, old documents and magazines. Most likely, due to many New England towns not having complete or in-depth town histories, some powder houses have unfortunately been omitted. Therefore, it is hoped that readers who know or learn of any powder house, powder magazine or gun house that has been left out would kindly inform the publisher so they can be included in an updated reprint of this work in the future.

I hope readers will enjoy learning about these fascinating forgotten yet historic New England landmarks and the significant roles they played in American history over the past 375 years.

NEW ENGLAND CITIES/TOWNS WITH POWDER HOUSES

Connecticut

Bridgeport, Fairfield, Groton, Hartford, New Haven, New London, Norwich, Stratford, Voluntown, Wallingford.

Maine

Bath, Belfast, Boothbay Harbor, Brunswick, Buckfield, Buxton, Calais, Durham, East Machias, Eastport, Edgecomb, Farmington, Gorham, Gray, Hallowell, Hollis, Kittery, Machiasport, Phillips, Phippsburg, Portland, Prospect, Saco, South Berwick, St. George, Union, Waldoboro, Westbrook, Wiscasset, York.

Massachusetts

Amesbury, Andover, Arlington, Ashburnham, Attleboro, Berlin, Beverly, Bolton, Boston, Bradford (also see Haverhill), Braintree (see Quincy),

Brookfield, Cambridgeport, Canton, Charlestown (also see Somerville), Chelsea, Concord, Dedham, Dover, Dracut, Duxbury, Easton, Essex, Fitchburg, Foxborough, Groton, Harvard, Harwich, Haverhill (also see Bradford), Hingham, Holliston, Ipswich, Kingston, Lunenburg, Lynn, Lynnfield, Manchester-by-the-Sea, Marblehead, Medfield, Medford, Medway, Middleton, Millbury, Milton, Needham, Newbury, Newburyport, Newton, Norton, North Brookfield (see Brookfield), Oxford, Peabody, Petersham, Pittsfield, Plymouth, Quincy, Rehoboth, Rowley, Roxbury, Salem, Salisbury (see Amesbury), Sherborn, Somerville (also see Charlestown), Southborough, Springfield, Sterling, Stow, Sudbury, Sutton, Taunton, Tewksbury, Topsfield, Walpole, Warwick, Watertown, Wayland, West Bridgewater, Westborough, Westminster, Weston, Westport, Woburn, Worcester.

New Hampshire

Amherst, Bedford, Canaan, Candia, Charlestown, Concord, Dover, Exeter, Hampton Falls, Hanover, Haverhill, New Castle, Plainfield, Portsmouth, Rochester, Rye, Sanbornton.

Rhode Island

Bristol, Newport, Providence, Tiverton, Warren.

Vermont

Bennington, Castleton, Corinth, Newbury, Orwell, Pittsford, Rutland, Shaftsbury.

CHAPTER 1
HISTORY AND PURPOSE OF POWDER HOUSES

For anyone who thoroughly enjoys reading and studying early American history, New England must truly represent the closest thing to historical paradise this side of heaven. Here the history enthusiast can easily find a wide array of fascinating historical landmarks, people, places and events that date back into the early 1600s. There are a seemingly endless number of ancient landmarks that dot the scenic rolling hills and valleys of New England. These landmarks provide quiet testimony and visual reminders of the rich heritage so closely associated with the birth of our nation and with the struggles of our early ancestors who had to settle and tame the once vast New England wilderness into thriving homesteads and close-knit communities.

Many of these familiar New England landmarks include white steeple churches; rustic stream-lined mills; old taverns and inns; imposing college, academy and university buildings; stately tree-lined town commons; tiny one-room schoolhouses; sprawling old graveyards; abandoned town pounds; white-painted bandstands; towering lighthouses; quaint country stores; endless miles of stonewalls; weather-worn garrison houses; historic forts; and stately old Georgian, Federal and Greek Revival–style homes. Unfortunately, missing among this long list of notable New England landmarks are the 201 known powder houses that once dotted the New England countryside, serving as vital military arsenals for many communities. These old forgotten powder houses played a pivotal and significant role in the establishment and preservation of early American Independence, and three of them unwittingly became home to the catalyst that started the American Revolution.

Most readers may be unaware that it was three particular powder houses—one located at Charlestown, now Somerville, Massachusetts; another in New Castle, New Hampshire; and the last in Concord, Massachusetts—that the British Army initially set in their sights. Soldiers were marching (or sailing) toward them for the purpose of confiscating all the powder, guns and ammunition stored within all three of these powder houses in 1774 and 1775. The British were doing this so that valuable gunpowder and arms would not fall into the hands of American "rebels" and thus be used against the British should war eventually break out within the unruly American colonies. On September 1, 1774, the British succeeded in capturing the gunpowder at the Charlestown powder house, causing great concern within the region with the "powder alarm." The powder alarm was the term used to warn town officials in New England communities to secure and protect their supplies of gunpowder from potentially being confiscated by the British Army. Fortunately the Americans were able to secure the firearms and gunpowder from each of the other two powder houses before the British were able to confiscate them. Little did the British realize that their march to the Concord powder house under the cover of darkness on the morning of April 19, 1775, would be the final incident that would actually trigger the start of the American Revolutionary War.

These powder houses—or, as the author prefers to call them, "Arsenals of American Independence"—served various communities as storehouses for military firearms, ammunition, artillery equipment and gunpowder in the event of a possible attack on the community or region. Through the years, these attacks could come in the form of Indians, British armies, British Tories or even Southern Confederates. Local militia companies also utilized the firearms and ammunitions stored within these powder houses for annual spring and autumn militia training exercises. These twice-yearly militia trainings were designed to keep all able-bodied male citizens between the ages of sixteen and fifty militarily prepared in the event that a war or battle should break out. Therefore, some of these old powder houses played a significant and useful role in helping to secure American independence. They also continued to help maintain that freedom and independence during the early years when the United States was still a young and fledgling nation.

In many New England towns where a militia company was maintained, the supply of powder, ammunition and armaments was stored in a shelter designed for that purpose. Most of these powder houses were made of brick and were usually square or circular in shape. A few were constructed of granite, and a few others were made of wood. These powder houses were usually constructed on a stone floor or on rock outcroppings atop small hills that were often locally known as "Powder House Hill." Most square powder houses were capped with a hip roof, while circular powder houses like Portsmouth, New Hampshire, Haverhill and Newburyport, Massachusetts all had conical roofs. Amesbury, Massachusetts had an unusual dome-styled roof. Eastport, Maine, and Fairfield, Connecticut were unusual in that they were designed as long stone or brick archway-shaped structures.

The powder house was often located with an eye to safety in most towns and to convenience in others. Those with an eye toward safety were usually constructed at some distance from the more settled sections of the community, such as Exeter, New Hampshire; and Ashburnham and Dedham, Massachusetts. Otherwise, powder houses were built near town commons or in a corner of a graveyard. Harvard, Massachusetts, was unusual in that it built a powder house quite close to the meetinghouse and town pound. Many New England communities also housed their stock of gunpowder and ammunition in the attics and cellars of meetinghouses, barns and, in one case, even a corn crib since there were no stoves or fireplaces in meetinghouses much before 1820. Sunday worshippers were known to flee meetinghouses during lightning storms in the event that lightning might strike the building and ignite the hidden supply of gunpowder.

Closely related to the powder house or powder magazine was the gun house. These structures were erected by the state in towns where the militia unit consisted of an artillery company. Gun houses were constructed on state-owned land and were not necessarily located in or near the center of town. In a few cases, some gun houses were built near homes or public buildings like Marblehead, Massachusetts, and Portsmouth, New Hampshire. Gun houses appear to have been built in their respective towns to suit the convenience of the local militia companies. The most recent gun house was built at Concord, Massachusetts, in 1960 to house equipment essential in re-enacting Revolutionary War battles.

Though not among the typical powder houses primarily recorded within this book, four commercial powder houses have also been included within this text due to the almost unimaginable destruction three of them visited upon their respective communities. The fourth commercial powder house is mentioned because it caused considerable fear to residents living within a Manchester, New Hampshire neighborhood in 1903 shortly after another commercial powder house in Tewksbury, Massachusetts, blew up, killing twenty-two people and injuring sixty to seventy others. The Tewksbury explosion also damaged hundreds of houses and businesses. With the exception of these four commercial powder houses, all other powder houses that were actually part of various gunpowder-making mills scattered throughout New England were not included, with the exception of Andover, Massachusetts, because gunpowder manufacturing is another separate topic of study.

Readers should keep in mind that the first battles and skirmishes of the American Revolution were fought over preventing the British from capturing the arms and ammunition stored at powder houses in New Castle, New Hampshire, on December 14 and 15, 1774, and at Concord, Massachusetts, on that eventful day of April 19, 1775, when the "shot heard round the world" took place at Concord's historic North Bridge. As it was, American Patriots deeply resented that the British had captured all the gunpowder, guns and ammunition stored at the towering stone powder house in Somerville, Massachusetts (then a part of Charlestown), on September 1, 1774, resulting in the historic event forever known as the "powder alarm" that quickly spread throughout the New England countryside warning Americans that the British had confiscated the vitally important military stores at Charlestown/Somerville. Therefore, because of the powder alarm, American Patriots were not about to allow future supplies of gunpowder, ammunition, firearms and cannons to fall into British hands again if they could possibly prevent it. Thus the reason many of these old powder houses are such important and overlooked portions of our American heritage. They—like other symbols of the American Revolution, War of 1812, and the Civil War—should be proudly preserved and cherished for centuries to come.

Paul Revere made his first "midnight" ride from Boston, Massachusetts to Portsmouth, New Hampshire, on the night of December 13, 1774, to warn locals that the British were planning to sail to New Castle, New Hampshire, to capture the ammunition stored there in the powder house at Fort William & Mary. Just four months later, on the night of April 18, 1775, Revere made a second "midnight" ride in which he attempted to warn locals between Boston and Concord, Massachusetts, of the British Army's march to capture the arms and ammunition stored at Concord in a wooden building serving as a powder house. He was captured by a British patrol before he could reach his Concord destination. Fortunately, a second rider by the name of Dr. Samuel Prescott successfully reached Concord in time to warn of the approaching British Army, and members of the Concord militia were able to remove or destroy most of the supplies in the powder house/store house and were able to prepare themselves to fight the British once they reached Concord. Within three weeks after the Battle of Lexington and Concord in mid-April 1775, powder houses were seized by American Patriots in

Connecticut, New York, Maryland, Virginia and Georgia, as was the vitally important cache of guns, cannons and ammunition captured by Ethan Allen from the powder house at Fort Ticonderoga, New York, on May 10, 1775. The munitions captured from Fort Ticonderoga were later used atop Dorchester Heights overlooking Boston Harbor in early 1776, which finally forced the British Navy to evacuate Boston, thus reinforcing again why these powder houses played such a significant and pivotal role in the formation of our nation.

Ironically, the majority of these powder houses were built in the first twenty-five years of the nineteenth century. Consequently, they ended up playing an instrumental role in helping to win America's second War of Independence—the War of 1812. The significance of this "forgotten war," fought for a second time against Great Britain, has unfortunately been largely overlooked by students of American history. It was this war that reaffirmed America's hard-fought goals of securing the ideals of freedom and democracy, which were successfully won during the American Revolutionary War of 1775–83. It was the winning of the War of 1812 that firmly secured America's right to be an independent republic capable of defending itself against belligerent nations that sought to undermine our strength and attack our borders. Winning the War of 1812 proved that the young and growing nation was indeed strong and determined enough to maintain its independence and sovereignty.

Upon defeating the British for a second time in 1815, the United States of America proved that a democratic republic was indeed a practical, viable and suitable alternative form of government versus the many totalitarian, monarchical or dictatorial forms of government that existed nearly worldwide at that time. After winning the war, the United States was now about to embark on a path that would eventually achieve manifest destiny from sea to shining sea, and thus along a path that would allow the United States to gradually become a strong and powerful leader of nations on the worldwide front at the turn of the twentieth century.

This book is the first comprehensive study of all the known powder houses erected throughout the six New England states. Many of these 201 old powder houses still stand either in forts, on lonely hilltops or in remote, isolated locations usually found just outside village or city limits. Many others have disappeared due to neglect. Sadly, these historic structures have largely been forgotten and overlooked by historians of New England history and architecture. Fortunately, in recent years there has been a resurgence of awareness and interest in restoring several existing powder houses.

Scenic view of the restored 1771 Exeter, New Hampshire powder house overlooking the Squamscott River. *Author's photo.*

Hopefully, the author did not overlook many communities that had or currently have powder houses and gun houses. Unfortunately, many town histories fail to record them, or some towns do not have town histories, eliminating a vital resource for learning of their existence. A search of the Internet found several others now mentioned in this book, but still a few have probably gone undetected. Diligent efforts were made over several years to locate every powder house in New England. Undoubtedly, there are also a few others that exist, or once existed, in many of the ancient forts that currently stand or formerly dotted the rocky seashores and wooded inland communities of New England.

CHAPTER 2

CONNECTICUT POWDER HOUSES

Connecticut—the Nutmeg State—joined the Union as the fifth state on January 9, 1788. At the time of the American Revolution (1775–83), it was the home of the famous American Patriot Nathan Hale, who was hanged in September 1776 for being caught carrying British military secrets to his American military superiors. On the other hand, Connecticut was also the birthplace of the infamous general Benedict Arnold, who, despite being a rising star in the American army, switched his allegiance over to the British during the Revolutionary War in September 1780 when he tried to hand over the fort at West Point, New York, to the British Army. He also ordered the British Army to burn Connecticut in 1781. To this day, the people of New London have neither forgotten nor forgiven this American traitor who ended up dying alone, distrusted and deservedly unappreciated in England on June 14, 1801—ironically, Flag Day here in America.

Connecticut suffered severely from the effects of the Revolutionary War, particularly during the Fort Griswold Massacre, where eight hundred British soldiers slaughtered eighty surrendering American Patriots during the Battle of Groton Heights, also known as the Battle of Fort Griswold, on September 6, 1781. New London, a port harboring a large number of American privateers and located across the Thames River from Groton, was set afire in September 1781 by the command of British general Benedict Arnold, who was born just up the road in nearby Norwich. Other Connecticut towns, such as Danbury, Fairfield, Greenwich, New Haven, Norwalk and

Ridgefield, suffered terrible destruction when attacked or burned by the British during the Revolutionary War.

Old New-Gate Prison, an abandoned 1707 copper mine located in East Granby, was transformed into a prison in 1773, and General George Washington later used this prison to confine British sympathizers known as Loyalists or Tories, as well as captured British troops, during the Revolutionary War. This must have been one of the most unbearable prisons in all of American history. One can only imagine the horror the prisoners felt as they descended down into the damp, dark, cold, rat-infested mineshaft and tunnels located forty-five feet below ground level. Old New-Gate Prison was closed permanently when Connecticut opened a new state prison at Wethersfield in 1827. This historic landmark is a must-see for American Revolutionary War enthusiasts.

The following is a listing of all the known powder houses, powder magazines and gun houses in Connecticut.

BRIDGEPORT

There was a powder house in Bridgeport as early as 1822 and perhaps much earlier, as it is mentioned in documents relating to the installation of a municipal wooden water pipe project in two sections of Bridgeport. No further information could be found about this structure, which no longer stands.

FAIRFIELD

The town voted to build a powder house on December 27, 1769, to "receive and lodge the town magazine." This long, arched brick and stone structure was completed in 1770 and is one of only two known powder houses in New England built in this design. It still stands on Unquowa Road, on the side of a hill in a section of Fairfield known as "Black Rocks." The land for it was donated by Levi Jennings. It stood about one mile from the seashore and halfway between Fort Union and Fort Defense, both used during the War of 1812, with the latter once standing at the Lower Wharf near the Mill River entrance, now known as Southport Harbor. Fairfield's powder house was also used during the War of 1812.

The 1770 Fairfield, Connecticut powder house is one of only two of this style still standing in New England. *Matthew E. Thomas collection.*

Fairfield, Connecticut powder house with a roof built over the structure in 1924. *Matthew E. Thomas collection.*

This powder house was restored in 1924 by the Eunice Dennie Burr Chapter of the Daughters of the American Revolution. A protective wooden roof was constructed over the top of the powder house at that time. It was again proudly restored in 2009 by the same local chapter of the Daughters of the American Revolution, which replaced both the old roofing and a wooden door. It also donated a bench for visitors to use and enjoy while visiting this historic powder house.

Groton

On the grounds of historic Fort Griswold stands a large and sturdy granite block powder house built in 1843, one of the youngest of New England's many powder houses. An earlier powder house was in existence within Fort Griswold in 1781 when it was attacked by the British during the Revolutionary War.

On September 6, 1781, Fort Griswold was the scene of a major Revolutionary War battle. Some 150 American Patriots held off a determined British attack force consisting of 800 troops. Utilizing effective cannon fire and the fort's strong stone walls to their advantage, the American defenders, under Colonel William Ledyard, inflicted serious casualties upon British troops. During the battle, the British temporarily fled the fort when they thought the powder house would explode. Upon resuming the fighting, the Americans were overtaken by a stampeding British bayonet charge. As Colonel Ledyard attempted to surrender his sword, he was run through by one of the British officers, encouraging the massacring of over eighty surrendering Americans—including those who were wounded and lying on the ground. This brutal display of cowardly bloodletting by the British (caused by confusing and uncertain signals during the battle) forever sealed American hatred toward the American traitor turned British general Benedict Arnold.

During the Battle of Fort Griswold in 1781, General Benedict Arnold ordered that a trail of gunpowder be laid between the barracks and the powder house and the gunpowder lit. A fire was set in the barracks, and the British retreated to nearby ships. As they sailed down the Thames River, they expected the fatal explosion, but nothing happened. At dusk, after watching the rear guard retreat, Major Peters of General Tyler's American military brigade approached the fort gate. Entering the gate, he rushed to

This 1843 granite-block powder house was within Fort Griswold at Groton, Connecticut. *Matthew E. Thomas collection.*

the well pump for water and extinguished the fire before the powder house could blow up.

This was the last important military engagement in the northern colonies during the Revolutionary War.

HARTFORD

Hartford, the capital city of Connecticut, had the unique distinction of having a temporary powder house contained in a two-story brick schoolhouse—a most dangerous combination at best.

The fatal accident known in the annals of Hartford as the "blowing up of the schoolhouse" occurred in the Town Square in a large brick schoolhouse built in 1758–59 that formerly stood in front of the old American Hotel. Friday, May 23, 1766, was the day set aside by the Connecticut legislature as a day of thanksgiving for the recent repeal of the much-hated Stamp Act imposed by Great Britain on November 1, 1765. The repeal caused overwhelming joy in the people, and May 23 was ushered in with the ringing of bells, firing of twenty-one cannons, displaying of colors on ships in the river and making of preparations for a grand illumination in the evening.

The state legislature had granted a quantity of powder for the festivities, which was promptly deposited in the first floor of the schoolhouse, which had been designated as a temporary powder house for the occasion.

A number of local young men had gathered together in an upper chamber of the brick schoolhouse to make sky rockets while the powder being stored in the room below was being distributed to the local militia. Two companies of soldiers had just received a pound of powder for each man, when the powder scattered by this delivery left a trail and was set afire by boys. In an instant, the schoolhouse/powder house exploded and was reduced to a heap of debris and ashes. Twenty-eight people were buried in its ruins, six of whom died after being taken out of the crumbling mass, and the others were more or less injured. Those killed in the explosion were Levi Jones, Dr. Nathaniel Ledyard and William Gardiner, all young and newly married men; Richard Lord, son of John Lord; John Knowles, an apprentice to Thomas Sloan, a local blacksmith; and Richard Burnham, aged nineteen, son of Elisha Burnham. This tragic event created much hysteria in the aftermath of the explosion.

According to the *Hartford Courant* of May 23, 1766, the injured included the following individuals:

> Samuel Talcott, Junior, *very much burnt in his face and arms.*
> James Tiley, *goldsmith, had one of his shoulders dislocated, and bruises on other parts of his body.*
> John Cook, Junior, *had his back and neck hurt much.*
> Ephraim Perry, *slightly wounded.*
> Thomas Forbes, *wounded in his head.*
> Daniel Butler, *(the tavern-keeper's son) had one of his ankles put out of joint.*
> ——— Wadsworth, *(Capt. Samuel Wadsworth's son), is much wounded and burnt, in his face, hands, and other parts of his body.*
> John Bunce, Junior, *wounded in the head.*
> Normand Morrison, *a lad that was a good deal burnt and bruised.*
> Roderick Lawrence, *slightly wounded.*
> William Skinner, *had both his thighs broken.*
> Timothy Phelps, *had the calf tore off from one of his legs.*
> Valentine Vaughn, *had his skull terribly broken.*
> Horace Seymour, *two sons of John Goodwin, a son of John Watson, and a son of Mr. Kellogg, were all slightly injured.*
> Two mulatto *and* two negro boys *were also wounded.*

NEW HAVEN

New Haven's powder house stood originally on the site of the Yale University Infirmary. It was erected about 1774 but no longer exists.

After the news of the Battle of Lexington and Concord reached New Haven on April 21, 1775, Captain Benedict Arnold and fifty of his company of the Governor's Foot Guard prepared to march to Boston. The New Haven selectmen refused to furnish Arnold with necessary gunpowder from the powder house. In response, Arnold defiantly threatened to break into the powder house and take what gunpowder he needed. The selectmen then immediately surrendered the powder house keys to him.

Today, New Haven celebrates a unique annual observance called Powder House Day. It commemorates the departure of Benedict Arnold's company of troops to Boston in April 1775. From 1905 to the present day, the scene has been reenacted each spring before crowds of spectators. Costumed militiamen march across the lower Green and draw up in formation in front of city hall. Officers march to the steps, where the mayor awaits them. Carefully rehearsed dialogue follows: "Captain Benedict Arnold and Lieutenant Jesse Leavenworth demand the powder house keys." These

View of the powder magazine behind the 1775 block house at New London, Connecticut. This is one of only two underground powder houses in New England, the other being at Fort Edgecomb, Maine. *Matthew E. Thomas collection.*

reenactors, along with other costumed officers, return to their troops, triumphantly carrying the keys to the powder house.

Despite Arnold's traitorous deeds later in the Revolutionary War, he is still depicted as a hero during the annual reenactment of this particular historical event in New Haven. No memorial to Benedict Arnold's memory has ever been built in Connecticut because of his unforgivable treason to his state and nation.

New London

The powder house at New London was built circa 1775 as part of the first Fort Trumbull. It was an underground powder house built into a knoll with a granite entryway. It stands behind the circa 1775 granite Block House. This was one of only two known underground powder houses in New England—the other being at Fort Edgecomb, Maine. This first fort was attacked during the Revolutionary War on September 6, 1781, by the British Army serving under the command of General Benedict Arnold, who turned traitor to America and ordered the burning of New London. The huge conflagration

The granite-block powder house at Fort Trumbull was built between 1839 and 1852 at New London, Connecticut. *Matthew E. Thomas collection.*

was started by the British when they unwittingly set fire to a storehouse full of gunpowder aboard the ship *Hannah*, which gradually ignited and destroyed over 140 New London houses, shops and public buildings.

A second powder house stands today as part of Fort Trumbull, which was rebuilt between 1839 and 1852. It has been designated a state park and is currently open to the public.

Norwich

A circa 1775 powder house, built of wood, formerly stood on top a small rocky eminence located behind the Congregational Church in the center of Norwich Village. During the Revolutionary War, this powder house was filled with several tons of gunpowder when a Tory (British sympathizer) was believed to have set it on fire. When the fire was discovered, it was just a little blaze, but attempting to put it out was deemed too hazardous an undertaking by those observing the event. Finally, a black man, induced by an offer of a considerable reward, made the attempt. As he was carrying water to extinguish the flames, after having gone about halfway up the hill, the building—with its large quantity of gunpowder—exploded, causing considerable damage to the many houses in the vicinity. The single explosion left scarcely a single vestige to show where the powder house formerly stood. The black man was left entirely unhurt by the explosion—most likely having been protected in part by the large granite rocks that made up the side of the hill that he was climbing.

Stratford

The powder house situated in the western portion of Stratford was located about three miles from the city of Bridgeport. Containing about one thousand kegs of powder and owned by the Union Metallic Cartridge Company, it exploded April 21, 1881, soon after 11:00 a.m. Buildings around the area were shaken as if an earthquake had taken place; people were thrown from their feet, and plate-glass windows were broken in many places. For a time, the wildest excitement prevailed. Shops were at once emptied of employees, with hundreds of women rushing wildly about the streets, crying. As it was supposed another explosion had occurred at the cartridge shop, all were hurrying in that direction. The workers at the cartridge shop were also wild with excitement, fearing a similar explosion to one that had occurred there a few months ago, and when the shop was shut down for the day, they left at once.

The damage to buildings throughout the city was estimated at $5,000. The explosion is believed to have been caused by the rifle shooting of two men in the vicinity of the powder house. One of these men, Egbert Pinney, standing about one hundred yards from the powder house, was badly cut about the head and the other badly shaken up but not otherwise injured. No lives were lost. Within the vicinity of the exploded powder house were two other structures, each containing about one thousand kegs of powder. The explosion was heard for miles around.

Voluntown

The only reference to a powder house in this town (or area) comes from the gravestone of a little girl named Happy Ann Kinne, born October 29, 1833, and killed on June 28, 1840, "during the explosion of a powder house." She was the daughter of Joel and Catherine (Church) Kinne and is buried in the Samuel Kinne Cemetery located behind the Connecticut State Sand Shed on Route 165 North, about a quarter mile on the southeast side of the intersection of Routes 201 and 165, just a few yards from the Griswold, Connecticut town border.

This information comes from Iva Arpin, who recorded the burials in this cemetery for the Town of Voluntown in 2003.

Wallingford

A large powder house was built in Wallingford circa 1776, much to the consternation of many residents during the outbreak of the American Revolutionary War. Guards were appointed to safeguard the powder house so that the despised British-sympathizing Tories would not find an opportunity to blow up the powder. This powder house no longer exists.

This view of the Fairfield, Connecticut 1770 powder house as it appeared in 1937. *Matthew E. Thomas collection.*

During the Revolutionary War, Wallingford served as a detention center for British Tories. The prisoners were confined to various houses throughout the town and were not allowed to travel about the town without a license from the head of the household where they were quartered.

CHAPTER 3
MAINE POWDER HOUSES

Maine—the Pine Tree State—joined the Union in 1820 as the twenty-third state. The movement to separate from Massachusetts and achieve statehood had started nearly forty years earlier. Maine has a rich and colorful heritage spanning over four hundred years. Explorers discovered this vast eastern land long before the Pilgrims landed at Plymouth, Massachusetts, in 1620, when it was covered with massive forests of white pine, spruce and hemlock trees overlooking a seemingly endless rocky coastline.

This land was occupied by several Abenaki Indians, a subdivision of the Algonquian nation, some of whom had dealt with early European explorers dating back to at least 1524, when Giovanni da Verrazzano documented the first case of a European actually laying eyes on Maine somewhere near Penobscot Bay. In 1607, a group of Englishmen known as the Popham Colony settled at Sagadahoc at the mouth of the Kennebec River, and it had every promise of being the first permanent English settlement in America. Unfortunately, two prominent key settlers died after just one winter, causing the other forty-three discouraged settlers to decide to return to England in 1608, thus abandoning the settlement completely and allowing Jamestown, Virginia, this honor in 1607.

From 1691 until March 15, 1820, Maine was a part of Massachusetts, separated by just eighteen miles of coastline that comprises a portion of New Hampshire, which, from 1643 until 1679, also fell under the jurisdiction of Massachusetts. Massachusetts had a law requiring each town under its jurisdiction to erect a powder house to contain each community's supply of guns and ammunition, including those in Maine, which explains why Maine has the second-largest number of powder houses of the six New England states.

With hundreds of miles of coastline, Maine was subjected to attack by British warships during the Revolutionary War. First, there was the seacoast town of Machias, Maine, which initiated on June 12, 1775 the first Revolutionary War naval battle at sea when locals on the small sloop *Unity* and the schooner *Falmouth Packet* captured the British man-of-war *Margaretta* after some close fighting that cost a total of five lives, including that of the *Margaretta*'s captain. Having captured the British ship, Machias men transferred the guns to the *Unity*, which they promptly renamed *Machias Liberty*.

Some 140 land miles to the southwest, Falmouth (now Portland), with its roughly two thousand residents, spent an uneasy summer in 1775 with war breaking out to the south and north of the town. Because of what had happened in Machias and at Lexington and Concord, Massachusetts, British Navy lieutenant Henry Mowat was ordered to attack and burn nine towns along the northern New England coast. He ignored eight of them but set his sights on Falmouth, and on October 18, 1775, he succeeded in attacking and burning three-quarters of the town, leaving 160 homeless families destitute during the winter and spring. This unexpected attack, meant to force the people of Falmouth and their sympathizers into submission, only served to strengthen their resolve toward independence from Great Britain.

Maine played a significant role in the outbreak of the Revolutionary War, and the Pine Tree State served as a theater for a few military actions of the war. In particular, it was the scene for one epic event.

That event was the March to Quebec, an expedition led by then colonel Benedict Arnold in 1775, which became one of the most dramatic episodes during the Revolutionary War. This march consisted of 1,100 American soldiers, mostly from New England, who marched with Colonel Arnold in September through the thick woods and swamps of Maine. The goal was to meet up with another expedition that marched up the Lake Champlain–Saint Lawrence route, in hopes of surprising the British at Montreal and then eventually uniting with Arnold's forces at British-held Quebec, where they planned to attack the fortified city in hopes of capturing Canada as a possible fourteenth American colony.

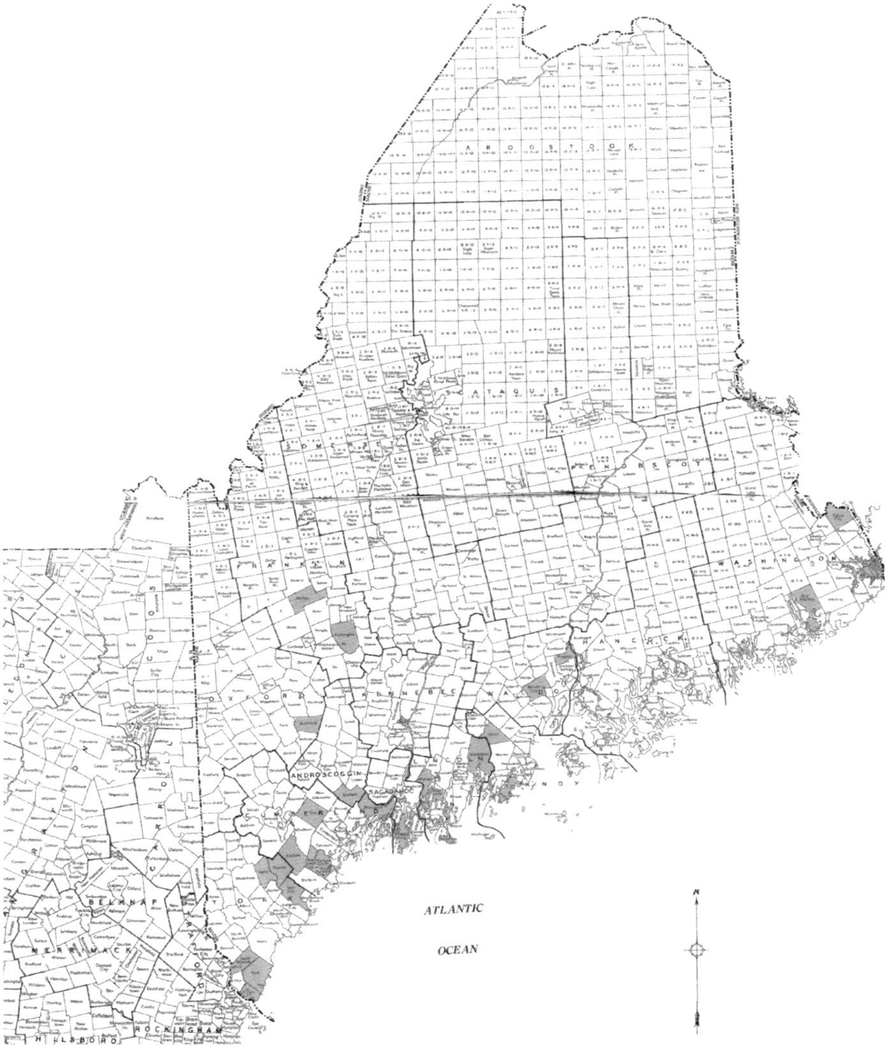

Map of Maine highlighting cities and towns with powder houses.

Arnold's greatly diminished and exhausted army of just six hundred emerged from the forests onto the shore of the Saint Lawrence opposite Quebec on November 8 after an excruciating march of roughly six weeks along the Kennebec River and through seemingly endless woods and swamps. Not until December 31, however, did the combined forces assault Quebec. The assault failed, with Colonel Arnold being wounded and General Richard Montgomery, commander of the other American army, fresh from capturing Montreal, killed on the Plains of Abraham outside Quebec. The British garrison beat back the Americans, killing many and capturing several hundred. Arnold kept Quebec under siege for a few months, but in the summer of 1776, his American army abandoned Canada forever. The plan to bring the somewhat sympathetic French into the American fold had failed. The terrible march through the dense Maine woods had all been in vain.

Maine suffered extensively from President Thomas Jefferson's Embargo of 1807–09. The embargo was in response to British infringements on American sovereignty and greatly incapacitated Maine's ability to trade with England, therefore hurting countless merchants and businesses. Nearly 60 percent of Maine residents living in its many seacoast communities were unemployed in 1809. The situation was so serious that soup kitchens were established in Portland to help its destitute citizenry. Despite this difficult time, Maine still supported "Mr. Madison's War" in 1812, though not completely, as many Mainers mounted considerable opposition, and smuggling in eastern Maine became practically epidemic.

During the War of 1812, Maine experienced minor naval battles and extensive smuggling off its coast and border with Canada. Much as America had fought for independence from Britain twice, the District of Maine also started thinking about its own independence from Massachusetts during and after the War of 1812. That independence was finally achieved on March 15, 1820.

The following are all the known powder houses, powder magazines and gun houses in Maine.

Bath

A gun house was built circa 1814–19 in Bath on the southeast corner of High and South Streets for the local Artillery Company, which was formed in 1795. There is no information to verify if the building still stands.

Belfast

All that is known about Belfast's powder house is that one existed, because a town ordinance restricted the quantity of gunpowder for sale in any place outside of the powder house to twenty-five pounds. There is no mention in the town history of when the building was constructed or how long it stood.

Boothbay Harbor

Boothbay Harbor built a gun house at the time of the War of 1812 on Eastern Avenue. The state of Maine sold the building in 1851 and it was converted into a carriage house. The structure was later moved to Montgomery Street, where it still stands.

Brunswick

Brunswick built its first powder house in 1804–05. It was constructed by Samuel Melcher, who agreed to build it for $20.00. In 1805, it was examined by a town committee and declared by them to be worth only $4.50. This amount the town paid, but it would not accept the building. In 1816, the town voted to build another powder house off Pleasant Street, on top of Powder House Hill. This brick structure amazingly cost $150.00—quite a difference from the $4.50 spent on the first powder house constructed just eleven years earlier! The 1816 powder house was sold by town vote to Mr. Samuel Haywood circa 1841–46. He sold the bricks to parties wanting them, leaving the structure to fall to the ground.

During the Revolutionary War, Brunswick partitioned off a section of the east end of its 1735 meetinghouse, near the roof, which was used as a powder magazine. There was never a stove or fireplace in this meetinghouse, which looked similar to the Rocky Hill Meeting House in Amesbury, Massachusetts. This meetinghouse was considered a safe place to store the town's supply of ammunition. It was later destroyed by fire in 1834.

A gun house was built in 1808 on Center Street. It was destroyed by fire in 1809. Another gun house was built immediately afterward and was used by the Brunswick Artillery for many years. It was later turned into a private dwelling on the southern side of Franklin Street near its eastern end.

BUCKFIELD

The Buckfield powder house was built in 1818 at a cost of thirty-five dollars. It measured eight feet square, with walls seven feet high. There are no traces remaining of this powder house.

BUXTON

Residents of Buxton voted on March 12, 1812, to build a brick powder house for the storage of "ammunition and arms for the Town's defense in case of War." Thomas Elwell was the lowest bidder and won the contract to build the structure for fifty-nine dollars. The building measures about six feet square and stands eight and a half feet high. The walls are twelve inches thick, and it has a shingled hip roof, formerly capped off with an impressive rooster weathervane that has been stolen twice. A door with ornate stout hinges is located close to the left-hand side of the building rather than being centered. This powder house stands at the end of a path about 250 yards off Route 22. This landmark was built in a large field that has reverted back to forestland in recent years. Elwell completed the project in the autumn of 1813. The following spring, gunpowder, lead balls, flints and camp kettles were stored in this little brick arsenal under the supervision of Captain Samuel Dunnell.

A nice granite stone marker with a bronze tablet thereon stands off to the side of this landmark that is on the National Register of Historic Places.

The town of Buxton has taken great pride in maintaining this wonderful old landmark, which is one of just three War of 1812–era powder houses still standing in Maine.

This 1812 brick powder house, with its tall rooster weathervane, which has been stolen twice, is located in Buxton, Maine. *Author's photo.*

Calais

A resolve issued by the Maine State Legislature in 1832 mentions the sum of $200 used to build gun houses in Eastport and Calais. On March 29, 1837, there was another resolve issued by the state to "sell the Gun Houses in East Machias and Calais and trade lots if possible in each case." Repairs to the Calais Gun House were done in 1843 and 1844 for $75 and $100, respectively. No other information could be found.

DURHAM

In 1812, William Webster and Barnabas Strout constructed a round brick powder house for the Town of Durham at a cost of seventy dollars. It stood near the old North Meeting House. Durham's powder house was destroyed by fire in 1879. The fire also killed Deborah Parker, whose home was located nearby the powder house.

EAST MACHIAS

The gun house in East Machias, built circa 1832–37, was authorized by the State of Maine to be sold on March 29, 1837. No further information could be found on this gun house.

EASTPORT

This is one of only two long-arched powder houses still standing in New England. It looked like the powder house in Fairfield, Connecticut. A small portion of this powder house still stands and is located behind the Shead Memorial High School. This powder house was built as part of Fort Sullivan about 1809, on top of Clark's Hill.

During the War of 1812, the fort was occupied by a small detachment of Americans. On July 11, 1814, a large British naval force under the leadership of Admiral Hardy appeared in the bay and immediately took over the fort. The British occupied the property until June 30, 1818, when diplomatic negotiations finally resolved that the fort was part of the United States and not part of the British Dominion of Canada.

The 1812 Eastport, Maine powder house is one of only two of this style still standing in New England; the other is in Fairfield, Connecticut. *Matthew E. Thomas collection.*

EDGECOMB

This historic old octagonal fort overlooks the Sheepscot and Back Rivers in the southern Maine coastal town of Edgecomb. Built circa 1808–09, it witnessed considerable action during the War of 1812 during many "alarms" that caused a militia to be stationed at the fort to protect it from menacing British war ships.

A brick and stone powder magazine was constructed near the river's edge and beneath the waterfront battery of this fort in the east bastion, where it measured six feet in width, seven feet high and seven feet long. The floor of the powder magazine is sixteen feet below the top of the bastion, which is five feet thick. The passageway leading down to the powder

Historic Fort Edgecomb in Edgecomb, Maine. The powder house that formerly stood in the immediate foreground of this photo was one of only two known underground powder houses in New England, with the other being in New London, Connecticut. *Author's photo.*

magazine was three and a half feet wide and ten feet long. Remnants of the powder magazine have recently been filled in to prevent injury of visitors to the fort. The entrance leading down to the old powder magazine was photographed in 1949.

Farmington

Farmington's powder house, built 1817, was a substantial brick building located on Powder House Hill. Sometime between 1846 and 1875, malicious boys of the town destroyed the powder house. There are many postcard views taken off Powder House Hill in Farmington. Unfortunately, there are none that show the town's former powder house.

Gorham

About 1798, Gorham built a powder house on the eastern side of Fort Hill Road, nearly opposite the mouth of Pendleton Lane. As a safety precaution, the brick powder house was placed several feet back from the road on a bare rock ledge. No trace of Gorham's powder house remains.

Gray

The Gray powder house was built circa 1812. It still stands as a converted hearse house and tool shed in the rear of the cemetery at Gray Corner.

A gun house was built at North Gray, a section of Gray. It was later moved and converted into a blacksmith shop. There is no information available on when it was built or whether it still stands.

Hallowell

Hallowell is located along the banks of the Kennebec River near Augusta, the capital of Maine. The first mention of erecting a powder house in Hallowell records appears on February 24, 1819, as the eleventh article on the town warrant: "To see if the town will give directions for building a Magazine for the safe keeping of Powder." This article was approved by voters at the March 10, 1819 town meeting, when the "selectmen were authorized to provide a suitable place for the safekeeping of powder."

This circa 1812 Gray, Maine wooden powder house was turned into a hearse house. *Author's photo.*

This view of the Hallowell, Maine brick powder house shows the "Thunder Jug" cannon in the background. *Matthew E. Thomas collection.*

The powder house was built on land sold to the town by Ebenezer T. Warren for thirty dollars on June 19, 1820. The powder house was already built at the time of this land transaction.

The final town record that mentioned the powder house was on March 8, 1820, when the expenses for the past year included "erecting a powder house at the cost of $145.00."

The powder house is a brick structure measuring 14 feet 2 inches by 12 feet 1 inch and 9 feet 4 inches high. The structure has gable ends and is roofed with cedar shingles. It sits on a small plot of land that measures 80.15 feet on High Street and runs 125 feet deep and 75 feet at the rear.

By 1947, the powder house was in serious need of restoration, and the property with the powder house thereon was gifted by the City of Hallowell to the Mary Kelton Dummer Chapter, Daughters of the American Revolution for one dollar. The chapter succeeded in restoring the building and placed a plaque on the building in 1948. It was again restored in 2003 by the chapter after being placed on the National Register of Historic Places in 2002 and qualifying for matching federal funds.

An unusual feature that is still intact with this powder house is the interior shelves that still have the names of several Hallowell male residents who stored their powder within the building circa 1837 and 1838 inscribed on them.

Interestingly, as late as 1871, the City of Hallowell revised its ordinances. Chapter 14 deals with the safekeeping of gunpowder. The powder house was deemed to be the depository for the safekeeping of gunpowder, and no person not licensed to keep and sell gunpowder was to have more than one pound of gunpowder in his home, shop or dwelling. A person licensed to keep or sell gunpowder was not to have more than twenty-five pounds at any one time, and gunpowder was to be secured in tin or copper vessels, in a situation where it could be readily removed in case of fire. Also spelled out were regulations for any floating vessel having gunpowder on board and the authorizing of the mayor or aldermen to search buildings for concealed gunpowder.

Another interesting landmark situated within a few feet of the Hallowell Powder House is the old Beeman Cannon, also known locally as the "Thunder Jug." Captain John Beeman, an enthusiastic, patriotic citizen of Hallowell, heard of the availability of the one surviving cannon out of fourteen that were taken from the British brig *Boxer*, which was captured by the U.S. brig *Enterprise* in September 1813. He elicited the assistance of some locals who together successfully won the cannon for Hallowell in 1839, with Captain Beeman funding the lion's share of the cost of the gun; thus it is

Interior view of original shelving of Hallowell, Maine powder house. *Photo courtesy of Sharon Ann Paradis.*

Beeman Cannon was also known as "Thunder Jug." It was captured off the British ship *Boxer* in 1813 and brought to Hallowell in 1839. *Photo courtesy of Sharon Ann Paradis.*

also known as "Beeman's Gun." For many years, the cannon was used to fire a Fourth of July salute down at the bottom of Central Street near a liberty pole that was also generously erected by Captain Beeman.

When exactly "Thunder Jug" made its way to Powder House Hill is unknown. However, it currently has a permanent place within a few yards of the Hallowell Powder House. The barrel of the "Thunder Jug" has long since been plugged so that it can no longer sound a salute on the Fourth of July, but the cannon was recently painted and is now a brilliant showpiece of Hallowell's illustrious past.

Hollis

The Hollis powder house was built in 1812 in a field opposite the old Dennett Homestead on Dennett Road. No further information exists.

Kittery

There was at one time a white, conical-roofed powder house built of rough stones that formerly stood atop a small hill in what is now the 1800 Portsmouth Navy Yard. The navy yard, which still sits on a small sixty-acre island in the Piscataqua River, is located in Kittery, Maine.

The year when the Kittery powder house was built remains unknown, but it most likely was built circa 1800–10 when the navy yard was in its first decade of existence. What is known is that at some unspecified date, part of the hill on which the powder house sat was leveled, along with the powder house, to make way for a new barracks and parade ground. No other details have been found.

Within the historic confines of Fort McClary are two powder magazines.

The site of Fort McClary, located off U.S. Route 1, has been used for nearly 275 years as a military post to protect the approaches to the Piscataqua River. The fort received its name in honor of Epsom, New Hampshire native Major Andrew McClary, who died at the Revolutionary War's Battle of Bunker (Breed's) Hill in June 1775. This fort was manned by military soldiers from the time of the Revolutionary War until World War I. It stands today as one Maine's most important historic forts that

Left: The 1808 powder house/magazine at Fort McClary in Kittery, Maine. *Author's photo.*

Below: Map of Fort McClary. *Courtesy of Maine Dept. of Conservation, Parks & Recreation.*

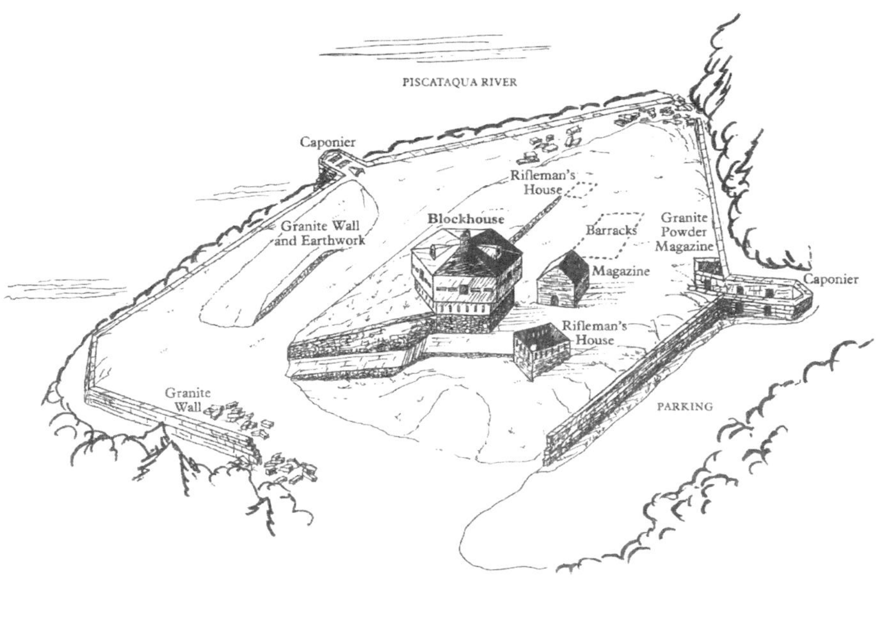

The 1864–68 granite-block powder magazine at Fort McClary. *Author's photo.*

preserves evidence of military history and changes in military architecture and technology.

The powder house in this fort dates back to circa 1808 and is the oldest structure still standing as part of this fortified site. The rectangular brick structure has reinforced brick walls and roof. There is a small window in the rear of the powder house directly opposite the entrance. There is currently no door hanging in the entranceway to this powder magazine. The entire structure measures approximately ten feet wide and eighteen feet long and has a shingled hip roof.

The second powder house in Fort McClary dates between 1864 and 1868, when Fort McClary was modified to meet the growing warfare technology during and after the Civil War. This powder magazine is made of thick granite blocks and low granite plastered hip roofs. It measures roughly sixteen feet wide and twenty feet long.

MACHIASPORT

Fort Machias, also known as Fort O'Brien, was built to defend against attacks from British naval forces. The O'Brien family of Machias was largely responsible for the building of the fort in 1775. A large garrison of American soldiers was stationed at Machias until late 1777. A small powder magazine was utilized on the grounds of this fort and was also used during the War of 1812 and the Civil War. Because Machias and Machiasport were prosperous lumber towns at the time of the Civil War, it was feared that the Confederate privateer *Alabama* or some other Confederate armed vessel might raid these prosperous communities. Therefore, a small military force was stationed at the fort for a time.

PHILLIPS

The Phillips powder house was located between the Parlin Road and the overhead Railroad Bridge in the Ross field on the right-hand side heading north. It was a strongly built wooden structure where a good supply of ammunition was kept along with firearms for an expected Indian attack. The building was provided with loopholes and situated on a suitable vantage point so that Indian tricks could be thwarted.

No dates are given in the town history that reveal the year of construction or year of destruction.

PHIPPSBURG

Fort Popham is a very historic two-tiered fort built in 1865 but never completed. Large granite blocks still litter the grounds where workmen left them while enlarging the fort. The fort, which overlooks the Kennebec River, contains four powder magazines and is located less than a mile from the Atlantic Ocean. This large granite fort is in the shape of a half-moon. This structure was built on the site of the old fort at Hunniwell's Point that dates between 1743 and 1748.

Portland

A powder house formerly stood on the site in 1936 of the Union Station. No further information could be found.

There is an existing powder house located in historic Fort Gorges, on Hog Island Ledge in Casco Bay outside of Portland, that was built in 1857–58. The fort was completed in 1865 and was constructed to provide additional military support for existing forts, including Fort Preble in South Portland and Fort Scammel built on nearby House Island in 1808. During a modernization project begun in 1869, sod-covered sand was added to Fort Gorges to protect gun encasements and powder magazines from attack.

Prospect

Fort Knox, located at 711 Fort Knox Road, is the most impressive fort north of Boston, Massachusetts. It was built in 1844 near a narrow channel of the Penobscot River and was not completed until 1869—four years after the end of the Civil War. The fort was utilized militarily up to the time of the Spanish-American War in 1898. This fort was named in honor of Major General Henry Knox, America's first secretary of war and commander of artillery during the American Revolution. General Knox lived in Thomaston, Maine (near the mouth of the Penobscot River), during the final years of his life.

During the American Revolution and the War of 1812, enemy British ships sailed up the Penobscot River, successfully attacking several vital river valley towns and claiming the area for the British crown. With America's eventual defeat of Great Britain during both wars, British claims to the area reverted back to the United States. By 1825, the United States government realized a defenseless Penobscot River could still be attacked and posed a serious threat to the thriving towns of Belfast, Bucksport and Bangor. Also, due to American and British tensions stemming from the boundary dispute over the location of Maine's northeast border during the 1840s, funds were finally appropriated in 1844 to build Fort Knox. This massive two-tier fort made of massive granite blocks contains four large powder houses inside the fort and another outside. All were used to store various military arms and ammunition. One of the powder magazines has been impressively restored to exhibit what the interior of a powder magazine actually looked like when the fort was in use

Interior view of one of five powder houses/magazines at the 1844–69 Fort Knox in Prospect, Maine. *Author's photo.*

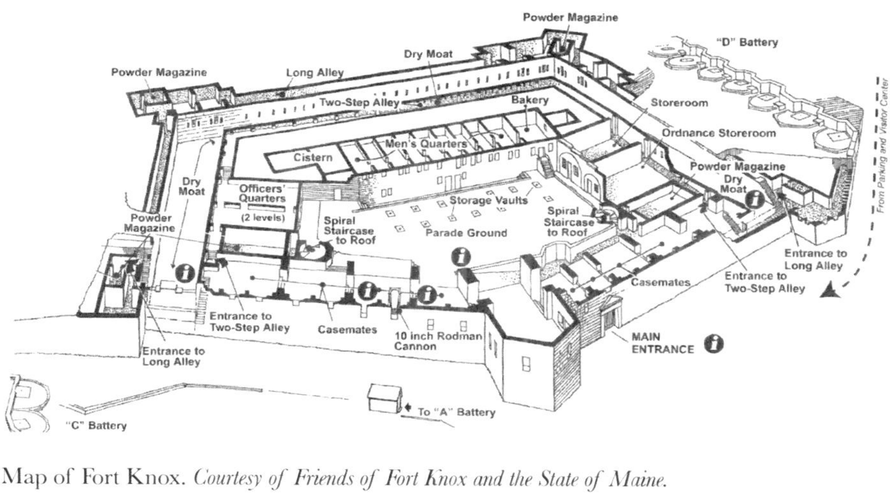

Map of Fort Knox. *Courtesy of Friends of Fort Knox and the State of Maine.*

during the 1850s. In addition to the five powder magazines, a large number of extensive granite- or brick-lined alley passages, spiral stairwells, men's and officer's sleeping quarters, a bakery, casemates, storage vaults and an ordnance storeroom all make for an exciting tour experience.

The circa 1858 shot furnace at Fort Knox was used to make cannonballs, which were later stored in the fort's several powder houses. *Author's photo.*

Historic Fort Knox overlooking the Penobscot River at Prospect, Maine. The fort contains five powder houses or magazines. *Author's photo.*

The fort also contains two brick shot furnaces, one built in 1858. Both were used to heat up large red-hot cannonballs, which were then fired from cannons, hopefully hitting wooden enemy ships and causing the eventual destruction of the ship by fire.

Fort Knox has undergone considerable renovations and restoration within the past ten years, thanks to the impressive determination of the Friends of Fort Knox, a nonprofit group that has partnered with the State of Maine in working to preserve and interpret this National Historic Landmark. The fort is open to the public May 1 to November 1 and is an outstanding landmark for military fort enthusiasts to study and explore.

Saco

All that is known about the Saco powder house is that it was built about the time of the War of 1812 and stood near the town common. It was later moved and turned into the house designed in the Victorian style that still stands. The author was unable to learn which Victorian house on either side of High Street near the corner of Beech Street is the correct building incorporating the original powder house.

South Berwick

This powder house was built in 1809, when South Berwick was still a part of Berwick. It wasn't separated and incorporated until 1814. It was built on Butler's Hill, later renamed Powder House Hill. All that is known of the structure is that it was set on fire by arsonists and completely destroyed on August 7, 1851, during the start of a destructive crime wave that greatly alarmed the residents of this old manufacturing mill town.

During the early years of the nineteenth century, militia musters had taken place twice a year for many years in South Berwick Village. This section of South Berwick was known as "the Plains" and was open countryside that was most suitable for such military trainings. Such was probably the reason a powder house was built in present-day South Berwick.

Another sad tragedy connected with Powder House Hill occurred on August 26, 1854, when two drunken, penniless transients came to South

Berwick and climbed Powder House Hill in the early afternoon, with only one coming down in the late afternoon. A month later, twenty-one-year-old Nathaniel Wadleigh came upon a decomposed body on Powder House Hill, clothed only in a shirt and lying near a marsh. The skull was smashed in, probably with a club or rock.

Since the two young men were transients, it was a difficult murder case to investigate. But the prosecutor was relentless in tracking down leads and facts, and soon it was discovered that William B. Smith was the guilty murderer of twenty-six-year-old Charles F. Brewster. The motive for the murder was robbery. Brewster had a five-dollar gold piece on his person, along with newly made pants and boots, which Smith took off Brewster after murdering him. Smith had made the mistake of swindling a man in central New Hampshire of twenty dollars and gave him the murder victim's pants and boots while inadvertently mentioning that he was heading to his home in northern Vermont. The swindled man reported the theft of twenty dollars, as well as the gift of pants and boots, to local authorities who in turned used the items as evidence to apprehend the murderer up in northern Vermont.

Smith was brought back to Maine, later convicted of murder and was sentenced to be hanged on January 25, 1855. He was sent to the Maine State Prison in Thomaston to await his death sentence. Death sentences were rarely carried out in Maine in those days, and Smith's death sentence was later commuted to life in prison in 1875. He presumably died at the Maine State Prison. Maine abolished the death penalty shortly thereafter in 1876.

ST. GEORGE

St. George is a small town located four miles east of Thomaston. The town is the home of old Fort St. George, built in 1809. That same year, a powder house that formerly stood within Fort St. George was built. The crescent-shaped fort was built on the rocky shore of the St. George River and was captured by the British in 1814 during the War of 1812. Now only the earthworks, cellars of the barracks, blockhouse and powder house remain as the only identifiable fort site in Knox County, Maine.

Union

Union's powder was stored in the garret of the old Meeting House. An unsuccessful attempt was made on May 8, 1815, to provide a powder house. On April 7, 1816, the town "voted to build a powder house, the expense not to exceed fifty-dollars." Accordingly, not long afterward, townspeople were surprised one morning at seeing a little wooden building about six feet square and ten feet high, with a peaked roof, perched on the highest part of the hill north of the town common. It had been made to order and hauled in the night to its elevated location. Here it remained for many years, and in 1851, when the town history was written by John Langdon Sibley, it "looked somewhat the worse for the storms and the boys' knives." An effort was made "in 1845 to have the town dispose of it and the utensils therein." Apparently it was left standing but has since disappeared after the 1850s.

Waldoboro

As early as 1817, the historic seacoast community of Waldoboro voted "to build a house for the safekeeping of the Town's Stock of Ammunition." At a later meeting, Charles Kaler, Gorham Parks and Payne Elwell were named

The 1820 Waldoboro, Maine powder house at Prock's Ledge. In *History of Old Broad Bay & Waldoboro, ME*, by Jasper J. Stahl.

to a committee "to draft a plan of a Powder House and to report at the next (town) meeting." However, the committee report given at a September 20, 1819 meeting was not acted on.

On May 8, 1820, $175 was appropriated for land. A suitable and safe site was purchased from Peter Prock and called "Prock's ledge."

In 1820, the circular brick powder house was erected. It measured fifteen feet high and seven feet in diameter, with a concave wooden roof. As the years passed and the need for a powder house diminished, Waldoboro's powder house fell into disrepair. In 1885, the building was dismantled and the bricks used to construct a new town building.

Wiscasset

During the War of 1812, Wiscasset residents recognized the need to build a place to store the town's stock of gunpowder.

At a town meeting held in May 1813, General David Payson, Colonel Ezekiel Cutter and Major Thomas McCrate were selected for a committee to recommend a suitable place to erect a powder house.

On June 9, 1813, acting upon the advice of the committee, the town fathers, according to the town history, "purchased from Jeduthan Boynton a small lot of land on Churchill Street twenty feet square lying a little northeasterly of the dwelling house of Michael Wharton, about 150 feet from the road." They also obtained "a perpetual right of way to pass, and re-pass, from the road with or without teams, for the inhabitants of Wiscasset."

Work on the circular brick powder house began in August 1813 and was completed in September of that year. Eight thousand bricks from Henry Cargill's brickyard cost $48.00. Nehemiah Somes was the brick mason and plasterer. John Warren, a local blacksmith, provided the hinges. The sturdy door was studded in every inch by bolt and nails, built to withstand nearly any form of attack. Phineas Kellam and Michael Wharton were paid for labor and materials used. On August 24, a charge of sixty cents for rum indicates the date the conical roof was raised. The total bills, including $20.00 paid for the land and $10.00 for Colonel Cutter's commissions on the job, came to $218.96.

Many years ago, the Wiscasset Fire Society restored and preserved this historic landmark. However, as of spring 2010, the grounds around the historic 1813 Wiscasset Powder House were sadly disappointing because

The 1813 Wiscasset, Maine powder house. *Author's photo.*

Postcard of the 1813 Wiscasset powder house with cows grazing nearby. *Matthew E. Thomas collection.*

the trees around the landmark have been recklessly cut down, leaving the property littered with cut lumber, shrubbery and massive roots. Early twentieth-century postcards depict a much more pleasant and scenic view, with cows grazing around the powder house that sits majestically on a high rocky hilltop, guarded dutifully by a "Lone Pine" standing behind this ancient landmark.

Over the years, Wiscasset has taken great pride in its powder house. Hopefully the site can one day soon be restored to its former serene and bucolic appearance.

YORK

The beautiful seaside town of York has a wooden powder house, measuring twenty-four feet by fifteen feet. It has a hip roof. The windows were most likely added after it ceased being a powder house. This powder house was originally built circa 1760 behind the present York Town Hall and then moved to the "Little Parade" in 1831 to serve as Charles O. Emerson's law office. Later, it was again moved to its present site at 276 York Street (formerly called "Scituate Men's Row" after four men from Scituate, Massachusetts, obtained land grants in York in 1642), where it was renovated and enlarged

Built circa 1760, the York, Maine powder house is the oldest in Maine. *Author's photo.*

with a kitchen/dining room addition. The building served as a real estate office for several years during the 1980s and mid-1990s. Since 1996, it has served as an art gallery/showroom operated by talented artist Jeanne Campbell.

According to another story handed down through generations, this powder house originally stood near the Preble Garrison House.

Just down the street from York's former powder house is another intriguing old landmark: the oldest public Gaol in America, built in 1653. This gambrel-roofed stone and brick structure is open to the public and takes the visitor back in time, exposing the various trials and tribulations of both prisoners and gaol (jail) keepers alike. This is a must-see for any early American history enthusiast.

CHAPTER 4
MASSACHUSETTS POWDER HOUSES

Massachusetts—the Bay State—was admitted to the Union on February 6, 1788, as the sixth state. It was here in Massachusetts where the second permanent settlement in America took place in 1620 (after Jamestown, Virginia, in 1607) when the Pilgrims landed at Provincetown Harbor on a cold, gray day in November 1620. After scouting the area, the Pilgrims shortly thereafter moved across the bay to Plymouth Harbor and began establishing their permanent settlement known as "Plimouth Plantation" in December.

Massachusetts has been called the "Birthplace of American Independence" and the "Cradle of American Liberty"—each rightfully so. It was from the determined efforts of such notable Revolutionary War–era Patriots as Sam Adams, Paul Revere, John Hancock, John Adams, Joseph Warren and James Otis that the revolt against suffocating British tyranny finally got a gradual foothold within the thirteen American colonies by the mid-1770s. After nearly a decade of imposed British taxes and restrictive acts—which included the 1764 Molasses Tax, 1764 Sugar Tax, 1765 Stamp Act, 1765 Quartering Act and 1773 Tea Act and culminated with the five Coercive or Intolerable Acts of 1774—all these oppressive laws combined to agitate and alienate the frustrated colonists. They gradually grew resentful at having little or no say in the British Parliament that voted on and approved these new taxes and laws without American representation. The colonists failed to see these acts as a reasonable means the British Parliament utilized to afford escalating costs of protecting the extensive thirteen colonies from Indian attacks and the still

Map of Massachusetts, Connecticut and Rhode Island highlighting cities and towns with powder houses.

unpredictable and troublesome French in Canada. (The British and colonists had only recently defeated the French in the Seven Years War of 1756–63, also known as the French and Indian War.) Instead, colonists saw these taxes as a deliberate attempt to tax and punish them into submission at the hands of a seemingly distant, unsympathetic and dictatorial British government led by that monarchical tyrant known as King George III.

One of the first hostile British acts against the American colonists took place on September 1, 1774, when the British Army, unbeknownst to local American Patriots, captured roughly 250 kegs of gunpowder stored within the massive old powder house located in present-day Somerville, which was a part of Charlestown until 1842. The powder belonged to several area towns, and the capture of it so alarmed the locals that they immediately organized a watch over all land and water routes to and from Boston. This event was known as the Powder Alarm.

On April 18 and 19, 1775, local Patriots living in Massachusetts towns located north and west of Boston got word that the British Army was marching westward from Boston to capture the arms and provisions stored in Concord in a building that served as a powder/store house (now the central part of the historic present-day Colonial Inn). They were not about to let a repeat of the Somerville incident happen again. Minutemen hid behind trees and stone walls between Lexington and Concord, fighting the British as they marched to and from Concord, with many Patriots participating in the historic battle at the North Bridge, where "the shot heard round the world" took place. To this day, no one knows for sure who fired the first shot on that fateful early morning of April 19 on Lexington Green that caused the deaths of eight militiamen and injured ten.

Soon, hundreds of determined minutemen from Massachusetts and southern New Hampshire were fighting at the Battle of Bunker (Breed's) Hill on June 17, 1775. It was a narrow British victory because their casualties were nearly twice those of the Americans, who retreated when they ran out of ammunition. The hard-fought battle emboldened the Americans, who realized that they, under some circumstances, could successfully fight against the most powerful army in the world.

Massachusetts, which included the District of Maine until 1820, was the first of the New England states to enact a law requiring each town in the commonwealth to erect a powder house. Though the vast majority of towns in Massachusetts ignored the law, it explains why Massachusetts and Maine have the largest number of existing or extinct powder houses of all the six New England states.

The following are all the known powder houses, powder magazines and gun houses in Massachusetts.

AMESBURY

According to the town history, "During the Revolutionary War and many years thereafter, Amesbury residents stored the town's stock of ammunition in Deacon Orlando Sargent's corn house." In 1802, it was deemed prudent "to build a powder house upon one of the "Pond Hills," also commonly known as "Captain Weed's Hill," away from nearby buildings." The town then built a small circular brick powder house, very similar to the existing Amesbury Powder House located atop Brown's Hill, off Madison and Monroe Streets. This 1802 powder house stood until the mid-1850s, when it was presumably torn down.

A second powder house was built on Brown's Hill, now called Powder House Hill, in the adjoining town of Salisbury in 1810 on land later annexed

This postcard of the Amesbury, Massachusetts 1810 powder house shows how it formerly appeared on an open hilltop. *Matthew E. Thomas collection.*

to the Town of Amesbury in 1888. This circular powder house, according to the Salisbury town history, "was built by Ephraim Weed, who was paid $10.40 for labor and hauling materials for the powder house. Isaac Whittier received $20.00 for labor and stuff which probably included the mason's labor. William Sawyer supplied lime for $12.00, and William Currier and Ebenezer Weed were paid a total of $22.50 for supplying bricks for the new structure. Total cost of the powder house was $64.90." The building stands roughly eleven feet high and has a conical roof. Around 1880, the powder house door was removed, the door space bricked in and the entire building, including the roof, was plastered over by the Town of Salisbury to prevent vandalism. In May 2010, the building was seriously vandalized, with someone trying to poke a hole into the side of the powder house and ripping out the bricks, plaster and concrete blocks therein. The debris littered the ground below the hole.

In 1998, the Amesbury Historic Commission erected a very nice historical bronze marker for the powder house at the bottom of Brown's Hill/Powder

This 2010 view of the vandalized Amesbury powder house shows it surrounded by trees. *Author's photo.*

House Hill on Madison Street, as well as constructed a stairway for visitors to more easily ascend the path leading up the hill to the powder house.

While on a visit to photograph this vandalized Amesbury landmark on June 7, 2010, it gratified this author to actually see a busload of schoolchildren making a pilgrimage up the hill with their teacher to visit the powder house. Hopefully, while witnessing the destruction of this landmark, the youngsters, ranging in age from eight to ten, were also taught by their teacher an important lesson for people of all ages: to not vandalize historical structures of any kind, as it diminishes our ability to learn firsthand from the unique landmarks of New England.

Andover

Little is known about the Andover powder house except that it was built circa 1775 or 1776 as part of the Andover Powder Mill—a major undertaking during the Revolutionary War started by Samuel Phillips, founder of the prestigious Phillips Andover Academy. Since there was a very severe shortage of gunpowder in America at the time of the Revolution, this newly constructed powder mill and powder house made an important contribution toward supplying much-needed gunpowder to the Patriot cause, thus justifying why these two powder houses warrant special inclusion within this book.

According to the diary of Mr. Philemon Chandler of Andover, under the date of June 2, 1778, "The Powder House at Andover was blown up. This disaster completely destroyed both the mill and powder house killing three men, and causing great consternation within the town."

The powder mill and powder house were rebuilt in 1778, but on October 17, 1796, a final explosion took place, destroying this second powder house and mill. Two men, including Mr. David Hall, were killed during the explosion of the powder house in Andover. The powder mill was never rebuilt after this disaster.

Both of Andover's powder houses were actually part of powder mills and not used specifically for storing the town's private stock of ammunition.

ARLINGTON

Arlington's brick powder house was built in 1811 on the left bank of Spring Valley. It stood near the brink of the hill where the valley empties into Spy Pond. This spot was chosen so that if an explosion occurred, the contents of the powder house would be blown into the pond. The structure no longer stands.

Arlington was originally incorporated in 1807 as "West Cambridge." The name was changed to Arlington in 1867. Prior to incorporation, present-day Arlington was part of historic Cambridge and was known as Menotomy.

ASHBURNHAM

According to Ezra Stearns's town history of Ashburnham, this structure stands "with a serene yet solemn countenance, overlooking the village of Ashburnham. Through the generations it has been a trustworthy custodian of the inflammable powder committed for safekeeping to its gloomy recesses."

The 1798 Ashburnham, Massachusetts powder house was built on a large boulder. *Author's photo.*

Town records state, "In 1798 it was voted by the town to build a powder house of brick, off Cushing Street upon Meeting House Hill, and to set it so as to convene Col. Jewett for the safekeeping of his powder, he giving ten dollars towards constructing the powder house. The town chose Esquire Wilder, Col. Jewett and Capt. Cushing a committee to build said house. Mr. Jewett died before the building was erected and Lieutenant John Gates was chosen to supply the vacancy. The building measures approximately 9 feet tall, 6 feet by 6 feet square, and is capped with a pointed four-sided shingled roof."

The Ashburnham Powder House stands on a large narrow granite boulder located on the side—not at the top—of Meeting House Hill. It is marked by a suitable sign and is proudly maintained by the town. The former twin-porch meetinghouse no longer stands at the top of the hill.

As the town historian Ezra Stearns remarked in 1886:

> *As a work of art, the powder house suggests no enthusiasm, and as an object of utility its loss could be easily supplied. But the village has become accustomed to its presence, and to all it remains a familiar object in the landscape. If removed the local associations of many years would be broken and the vision would rest upon the hillside unsatisfied. Like many objects and landmarks familiar to the recollections of youth, the powder house is not so fully prized in its continued presence as it would be deplored in its loss.*

ATTLEBOROUGH

On September 12, 1768, the residents of Attleborough approved the construction of a brick powder house in the Oldtown section of town near the end of Mt. Hope Road, not far from the Old Post Road. It was to be built on a small hill overlooking a plain that served as a training field for the local militia near the immediate vicinity of the First Congregational Church.

The powder house was built almost immediately in October and November 1768 and is circular in shape, measuring nearly twelve feet in diameter, and has a cone-shaped roof. Early records mention that six thousand bricks were used in its construction.

The building was constructed on land given for said purpose by Town Treasurer Jacob Newell in 1768. The following entry was made in the town book:

Above: Postcard of the 1768 Attleborough, Massachusetts powder house. *Matthew E. Thomas collection.*

Left: Photo of the Attleborough powder house with protective fencing. *Author's photo.*

Where as ye Town of Attleborough have agreed to Build a Powder house and set said house on ye High Hill eastardly from ye Meeting-House on my Land I give liberty to set said House there, and also for my Heirs and assigns I agree ye People shall have Liberty to Pass to and from said house to carry their powder or any other ammunition and bring off again ye same at any time or times so that they in Passing Regularly not to leave down fence or any Enclosing thing to Do Damage as witness my hand.
—Jacob Newell

The building is hardly noticeable from the road when foliage is on the trees. Not until a steep hill is negotiated on foot will the visitor get his first glimpse of this historic landmark.

The powder house was subjected to deterioration and vandalism over the years and was repaired in 1872 by concerned residents; the roof was re-shingled around 1938, and in 1948, restoration work to repair missing bricks and to secure the door with a new lock was authorized by the town selectmen. To the dismay of many longtime residents of Attleborough, selectmen also approved funding in 1948 to purchase fencing that would be placed around the structure.

Despite the fence, Attleborough residents now take great pride in having this unique landmark as an important part of their town's colonial heritage.

BERLIN

Near the center of town at the top of Powder House Hill, located off Route 62 East on a large outcropping of granite rocks, stands a small brick powder house built in 1814 during the War of 1812. It measures approximately five feet by five feet square and stands eight and a half feet tall, with a pointed, four-sided shingled roof. At one time during the early twentieth century, there was a flagpole attached to the west side of the Berlin powder house, and it was one of only two known powder houses in New England to have a flagpole attached to it—the other being the Somerville, Massachusetts powder house. This powder house is one of very few in New England whose entry door is not evenly centered with the structure but is instead constructed closer to the right front side of the building. The building committee consisted of Captain Henry Powers, Stephen Bailey and Captain Solomon Howe.

This is one of Berlin's most historic structures and has been admirably well maintained over time, despite years of graffiti attacks.

Postcard photo of 1814 Berlin, Massachusetts powder house circa 1910. *Matthew E. Thomas collection.*

Beverly

This historic North Shore town dates back to 1668 when it was incorporated. As early as 1727 the town stored all the town ammunition in the powder room of the cellar of the 1682 Second Congregational Church of Beverly. Because the town needed more space, Beverly voted to build its first powder house in 1767 on the south side of the Town Common. There is no record of what became of the 1767 powder house.

In 1808, Nathan Dane built an eight-sided powder house (the only eight-sided powder house still standing in New England) on a granite stone foundation, with a pointed eight-sided wooden shingled roof, making the

A 2012 photo of the 1808 Beverly, Massachusetts powder house. Note the remnants of white paint on the structure. *Author's photo.*

Circa 1920 postcard of the white-painted Beverly powder house. *Matthew E. Thomas collection.*

whole structure about nine and a half feet high. He built this structure on his own land atop Prospect Hill and sold it to the town of Beverly for thirty dollars. Polly Dane, his wife, was paid ten cents for forfeiting all claims to the land, which measured about 1,700 square feet. When this powder house was built, it stood in an isolated area. Today, this historic landmark looks somewhat out of place surrounded and almost hidden by a neighborhood of close homes. The Beverly powder house was put into use for the first time on April 16, 1808. As of 2013, the 205-year-old building has deteriorated and is in serious need of repair. This wonderful old landmark reflects one aspect of Beverly's very historic and colorful past. Hopefully the city or some patriotic organization will muster the necessary pride and financial funding in order to preserve this second-oldest public building in Beverly.

BOLTON

The four-sided brick powder house in Bolton sits atop a rocky outcropping of granite on Powder House Hill behind the present-day town hall. It measures about seven and a half feet tall, with a pointed, four-sided shingled roof. Though this small town was incorporated in 1738, it was not until circa 1812 when the town built its first powder house.

A 2010 photo of the Bolton, Massachusetts powder house. *Author's photo.*

Postcard of the Bolton powder house. *Matthew E. Thomas collection.*

Boston

Boston erected its first powder house atop a small hill near the Frog Pond on historic Boston Common in 1706. About the year 1769, just a few years before the start of the Revolutionary War, fortifications were thrown up at this site as a precaution againt potential war activities to come. This powder house was, according to the book *Days & Ways in Old Boston*, replaced with a new one in 1774 because the 1706 structure was located near the "Great Elm," also known as the "Liberty Tree," and was therefore more frequently exposed "to potential accidents on days of public rejoicing." This original

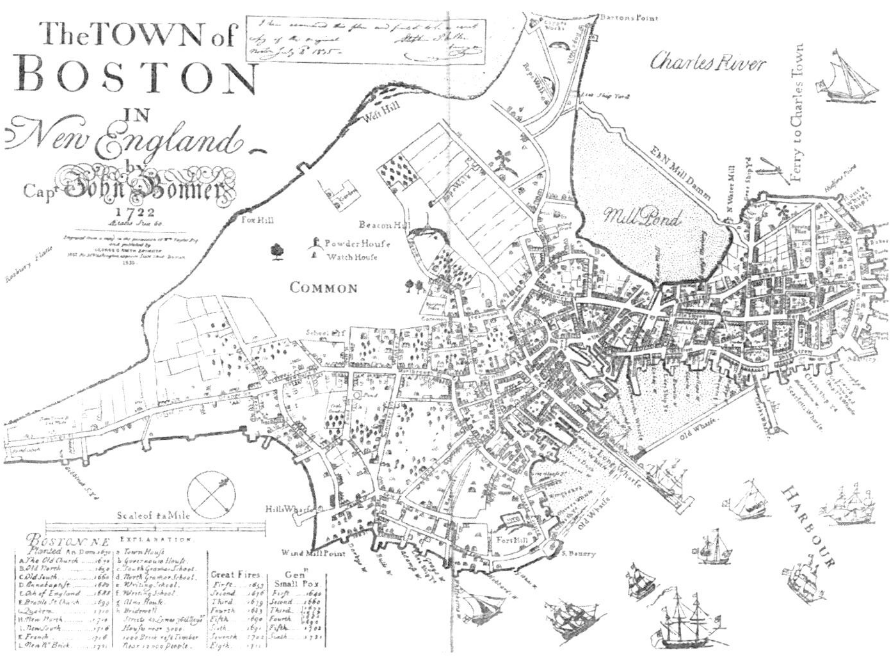

"1722 Map of Boston in New England," as drawn by Captain John Bonner. It shows the powder house and watch house on Boston Common as published in 1915 in *Days & Ways in Old Boston*, by William S. Rossiter.

powder house had a watch house about one hundred feet away from it, according to the 1722 map of Boston.

Between 7:00 and 8:00 p.m. on July 3, 1728, a tragic event took place near the 1706 powder house that was believed to be the first known fatal duel in America. Samuel Adams Drake explains in his book titled *Old Landmarks & Historic Personages of Boston*, "The combatants were two men of Boston: nineteen-year-old Benjamin Woodbridge and Henry Phillips. Supposedly either "cards & wine" were the cause of disagreement between the two, or "a woman was at the bottom of it." But whatever the cause, these two lads had long been friends, and something happened to stir up great enmity between them that only the sword could resolve. The two combatants dueled in the early evening sunset. When Woodbridge was stabbed by Phillips's sword under the right breast through to his back, the forefinger of his left hand was nearly cut off at the uppermost joint, supposedly caused by Woodbridge grasping the sharp sword while in great pain. Woodbridge, in return, had inflicted a light wound to Phillips's belly and fingers. Phillips, who immediately felt sorry for his actions, left to get help for his dying friend, but when he returned with a doctor, Woodbridge could not be found. It was supposed he had made it to his home or to some nearby house, but after they had knocked on several doors, Woodbridge still could not be found. It wasn't until 3:00 a.m. on the morning of July 4 that a search party found Woodbridge's dead body near the powder house, not far from where he had been stabbed. His body was carried to the house of Jonathan Sewell and buried on July 6 in the Granary Burying Yard. His funeral was attended by many influential members of the community, including the governor and members of the governor's council. In the meantime, Phillips fled Boston in the early morning dawn on the British man-o'-war *Sheerness* in fear of legal or vigilante reprisal. He escaped to Rochelle, France, where he was received by Jean Faneuil, the brother of Peter Faneuil of Faneuil Hall Marketplace fame. There Phillips died on or about May 20, 1729, less than one year after the unfortunate duel.

The duel, believed to have been the first in America in which a man was killed, caused the enactment of a new law in Massachusetts that decreed that any person fighting a duel "should be carried out publicly in a cart to the gallows, with a rope tied around his neck, and set upon the gallows for an hour, and then imprisoned for 12 months without bail." A person killed in a duel was to be denied "Christian burial" and interred "near the usual place of public execution with a stake driven through his body." Death was to be the penalty meted out to the survivor in a fatal duel.

A second powder house was erected in 1774 near the intersection of Pinckney and West Cedar Streets in Boston. This powder house was a very substantial structure made of Braintree, Massachusetts granite, with walls seven feet thick, and containing a bomb-proof arch. It was surrounded by palisades and was estimated to be able to hold one thousand barrels of gunpowder. There was also a watch house standing approximately one hundred feet east of this powder house.

In 1801, the Massachusetts state legislature ordered the demolition of the 1774 powder house. It was purchased by the Mount Vernon proprietors, who tore it down prior to 1804. There appears to be no further mention of powder houses in Boston proper after 1802.

A gun house was built at the foot of Boston Common in 1765. It was still in use in 1812 when the local military companies utilized it as their headquarters. In 1813, the park artillery paraded at the gun house on the Common "with their uniforms and side arms." In 1843, the gun house on the Common was removed a short distance to nearby Park Square. No information is known about its demise.

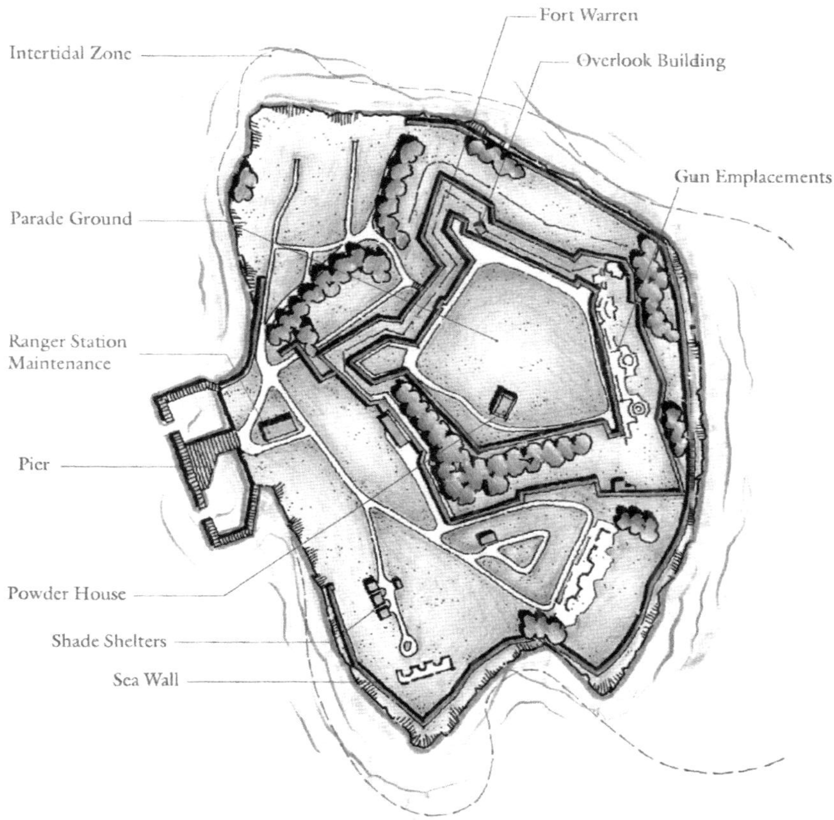

This map of George's Island in Boston Harbor shows Fort Warren, with the powder house in the center of the fort. *Courtesy of U.S. National Park System.*

A second gun house stood on Cooper Street at Boston's North End in 1863. On July 14, 1863, a great draft riot took place at this gun house when a mob attacked and conquered a small number of police who were attempting to serve draft papers to potential Civil War soldiers on Prince Street. The mob then proceeded to the gun house on nearby Cooper Street in an attempt to seize the cannon therein. They were thwarted by a small number of soldiers from Fort Warren who bravely guarded the gun house. Unfortunately, several soldiers were killed and wounded and the doors to the gun house nearly demolished when a general discharge from a cannon into the mob caused them to scatter and retreat to Dock Square. Eight rioters were killed. Some stragglers went to the Boston Common some distance away where others had assembled on hearing the news of the draft riot taking place. No information could be found to confirm its dates of construction or destruction.

BOSTON HARBOR POWDER HOUSES

Out in Boston Harbor on George's Island is the impressive Fort Warren, built in 1833 and completed during the Civil War (1861–65). It was named in honor of Dr. Joseph Warren, who was a prominent Revolutionary War hero killed during the Battle of Bunker (Breed's) Hill on June 17, 1775. The fort was constructed as a major fortification to protect Boston from enemy ships in future wars. The five-sided, pentagonal-shaped granite fort was originally designed to contain three hundred cannon and was one of the strongest coastal forts ever built in the United States. This historic fort, active during the Civil War, confined many prominent and not-so-prominent Confederate soldiers and governmental officers, including Alexander Stephens, the vice president of the Confederate States of America between May and October 1865. The fort saw military action up until the end of World War II.

Fort Warren originally contained two powder magazines, of which one survives. Both were presumably built in the first thirty years of its existence. The existing powder magazine measures approximately fifty-five feet long by thirty-six feet wide, has a gable roof and is surrounded by a large granite wall. A smaller powder magazine was formerly located on the north side of the fort. Fort Warren is now a National Historic Site and is a popular national park accessible by boat.

On Castle Island in Boston Harbor (which is connected to the mainland by an 1892 causeway) stood eight forts dating back to 1634. Since 1703, it was called Castle William after William III, King of England. Castle William was destroyed by the retreating British Army in 1775 as they evacuated Boston and was quickly repaired by American troops serving under Lieutenant Colonel Paul Revere. Castle William was also a refuge for British loyalists, including consignees of the tea, who fled to the castle after the 1773 Boston Tea Party. The name of the fort was changed to Fort Independence in 1799 when then president John Adams dedicated the site for a new five-bastioned brick fort, which was actually constructed between

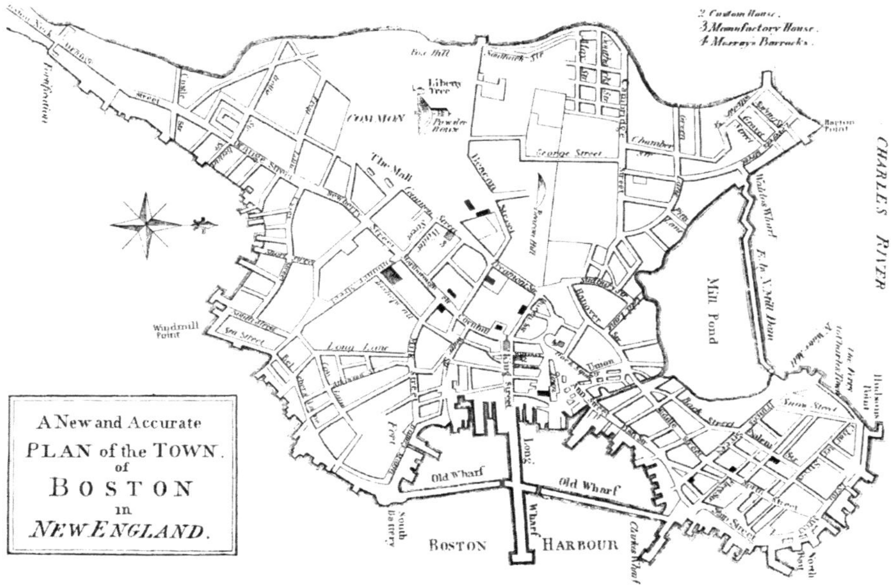

Circa 1769 map plan of the town of Boston. *Courtesy of New England Historical Research Associates.*

1801 and 1803 as the seventh fort on Castle Island. The current fort, built between 1834 and 1851, was larger and taller and was designed by Colonel Sylvanus Thayer, the "father" of West Point. It was constructed with granite blocks from quarries in Rockport, Massachusetts.

Fort Independence is the oldest continuously fortified site in the United States. Three powder houses once stood on the outside of the pentagonal five-bastioned fort, which was used during the Civil War as a testing ground for new cannons and armaments. The fort sits on twenty-two acres and has been a National Historic Landmark since 1970.

Bradford

All that is known about the Bradford powder house is that one was erected upon "Indian Hill."

BRAINTREE

See Quincy.

BROOKFIELD

A brick powder house measuring ten feet in diameter formerly stood near the Brookfield Town Pound when it was built in 1770.

After the Revolutionary War, the gunpowder owned by Brookfield was divided up among the three parishes and stored under the pulpits of three meetinghouses. North Brookfield's share remained under their pulpit until 1815, when on one Sunday afternoon, a terrific thunderstorm occurred, and a heavy bolt of lightning struck a rock just in back of the meetinghouse, and parted, one streak running under the church. The residents took the hint and removed the powder to a less public place.

No further information could be found as to what happened to the Brookfield powder house.

CAMBRIDGE

In 1817, the town of Cambridge sold "Captain's Island" to the State of Massachusetts, which in turn erected a powder magazine in 1818 on the island. Shortly thereafter, Magazine Street was built and ran from the heart of the city down directly to the powder magazine. The area where the powder house stands is known today as Magazine Beach. The state sold the property in 1884, and in 1886, the powder magazine was being leased from the new owners to the Pope Manufacturing Co. The powder house still stands in need of restoration.

CANTON

After the Revolutionary War, Canton's supply of gunpowder and ammunition was stored in Elijah Dunbar's chaise house. By 1807, the

The 1809 Canton, Massachusetts wood-shingled powder house. From *History of Canton, MA*, by Daniel T.V. Huntoon (1893) .

issue of Canton building a powder house was discussed at town meeting. A committee was appointed in 1809 to select a suitable site and to ascertain the expense of building a powder house. Later that year, the town built a powder house on an unnamed hill. It was constructed of stout planks covered with boards. It was shingled and crowned with a hip roof. It measured seven feet square and seven feet high and was fastened to the rock foundation with iron bolts. It remained at this location until 1858, when it was removed to the Town Farm.

Charlestown

In 1794, by an Act of Congress, the Charlestown Arsenal was established to supply weapons and ammunition to the U.S. Army and various organized militia in the New England states. There was a powder house on-site to contain whatever gunpowder or ammunition was needed for this purpose. After the War of 1812, the United States had increased its involvement on the seas, which necessitated the need for a naval repair yard in Boston Harbor. Charlestown was better suited for this endeavor, and the Charlestown Arsenal was turned over to the U.S. Navy while the army looked elsewhere for a new arsenal, which it established in nearby Watertown, Massachusetts, between 1816 and 1819. Watertown was chosen because it offered proximity to Boston, had available building materials, was easily accessible by water and had a higher elevation that made it easier to defend from potential enemy attack.

Charlestown was officially annexed to the City of Boston in 1874. It is the site of the historic 221-foot Bunker Hill Obelisk Monument, which was started in 1827 and dedicated on June 17, 1843.

Also see Somerville.

Chelsea

Though a powder house was built in 1811, no further information is available.

Concord

Ironically, of all the New England powder houses, the Concord powder house (actually now a portion of the present-day 1716 Colonial Inn on Monument Street) is the most illustrious structure of its type. It was to this particular powder house that the British Army was secretly marching (by way of Lexington) during the historic night of April 18 and morning of April 19, 1775, to prevent these vital munitions from falling into the hands of an increasingly growing faction of belligerent American Patriots. As a result, with the British Army secretly trying to march to

A portion of this section of the historic Colonial Inn at Concord, Massachusetts, housed the original powder house that the British Army was marching to on the fateful days of April 18 and 19, 1775, to capture the powder held by American patriots. This was the final straw that caused the actual outbreak of the American Revolutionary War, which lasted from April 1775 until the American victory at the Battle of Yorktown, Virginia, in October 1781. *Author's photo.*

This interior is a great example of what most early New England powder houses would have looked like with cannons, cannonballs and barrels of gunpowder stored within. This is a photo taken by the author at the Concord, Massachusetts National Historic Park and Information Center.

Concord's 1716 powder/store house, the "shot heard round the world" was fired, and the Revolutionary War began in earnest.

The immortal poem penned so eloquently by Henry Wadsworth Longfellow and titled "The Midnight Ride of Paul Revere" unfortunately fails to mention that Revere was actually captured *before* he reached Concord to warn that the British were coming to seize the town's supply of gunpowder, military stores and ammunition. Instead, Revere was just the better known of three Patriot dispatchers, the others being William Dawes and Dr. Samuel Prescott. Their purpose was to ride to Concord and warn of the impending arrival of the British Army. British regulars on the road to Concord stopped Revere, who was captured and later released. They tried to capture Dawes, who also escaped from the British and headed back to Lexington, and fortunately for American history, Dr. Samuel Prescott was able to reach Concord and nearby Acton to warn town officials of the approaching British Army. Only a small portion of Concord's stock of gunpowder and military stores was actually captured by the British, while some of it, along with flour, was thrown into the Concord village mill pond. But the majority of the munitions were hastily dispersed throughout the town and region by local Patriots for safekeeping from the British.

The Concord town history fails to mention anything about the town's second powder house, but town records of May 15, 1786, mention voter approval of building "a small house at the easterly corner of the Burying place upon the hill (to be) seven feet high and seven feet square in order to deposit the Powder and other Town stock in the future—and that the selectmen be a committee for that purpose." Presumably, this 1786 powder house was removed and replaced with the existing 1810 powder house.

A small square brick building still stands within the historic Hill Burying Ground atop the east end of the hill, but the measurements do not conform to the dimensions of the 1786 structure. This 1810 powder house measures about five feet square and seven feet high, with a pointed, four-sided, shingled roof formerly capped with a finial. Whatever the case, the building was restored in 2006 and re-roofed and is proudly maintained by the Town of Concord.

The citizens of the Town of Concord raised funds in 1960 to construct an authentic-looking gun house on Lexington Road to store the cannons, wagons, carriages and other military armaments used in the various Revolutionary War reenactments that take place on a regular basis in

Circa 1910 postcard of the 1810 Concord, Massachusetts brick powder house, still standing in the Hill Burying Ground, showing its roof finial. This was the third powder house built in Concord. The first was part of the Colonial Inn, and the second, built in 1786, was replaced by the 1810 structure. *Matthew E. Thomas collection.*

This 2012 photo of the 1810 Concord powder house shows it missing its roof finial. *Author's photo.*

historic Concord. The two-door brick structure with large wooden doors and long black iron hinges is the newest gun house in the state and is well maintained. It looks similar to the Marblehead, Massachusetts 1810 gun house. The Concord Gun House serves as a proud symbol of how dedicated and committed the citizens of Concord are to cherishing and preserving their community's rich and colorful military heritage.

Above: New England's newest gun house was built in Concord, Massachusetts, in 1960. Gun houses served a similar purpose as powder houses and powder magazines and were often used and maintained by local artillery companies. These structures were often large enough to house wagons, cannons, ammunition and other military apparatus. *Author's photo.*

Left: A sign for the 1960 gun house, maintained by the 1804 Concord Independent Battery Association. *Author's photo.*

DEDHAM

The South Shore community of Dedham (south of Boston) has the second-oldest record of building a powder house in New England. As early as 1653, the residents of the town paid Eleazer Lusher £1.5.8 old currency for work on building an "ammunition house." In 1673, the new meetinghouse was used for the storage of ammunition as the General Court of Massachusetts had ordered the town to prepare for war with the Indians.

Eighty-nine years later, in 1762, town records state that "it was put to vote to see if the Town will build a Powder House." That year it was "voted to have a powder house built on a great rock on Aaron Fuller's land near the Charles River." However, nothing was done until the May 1765 town meeting. Townspeople voted to erect a powder house. It was to measure "8 Feet Square on the outside, and 6 Feet High under the eaves, the material to be of lime, mortar, and brick."

Captain Eliphalet Fales, Daniel Gay, Ebenezer Kingsbury, Deacon Nathaniel Kingsbury and Captain Daniel Fuller were members of the building committee.

The powder house was erected by Daniel Fuller in 1766 on the top of a small rocky hill off Ames Street, now called Powder House Rock, on land owned by Aaron Fuller. The total cost of building the powder house was £12.6.8. It was used for the first time during the spring of 1767, when the building was stocked with ammunition, including a keg of flints, in readiness for the old flintlock guns. The town owned the powder house but not the land on which it sat. The first section was obtained in 1912 and the second in 1927.

A 1934 photo of the 1766 Dedham, Massachusetts powder house. *Matthew E. Thomas collection.*

Rare circa 1875 stereoview of the 1766 Dedham powder house situated atop Powder House Rock. *Matthew E. Thomas collection.*

By 1859, the powder house was in sad need of repair. The town attempted to tear down the powder house, but due to strong opposition, the building was saved and repaired by donations.

In 1886, it was necessary to repair the partially dilapidated building again. Work done adhered to the original design as closely as possible. A bronze marker was inserted in the front wall in 1886, bearing the following inscription:

The Powder House
Built by the Town
1766

Restoration work was again performed on the building in 1979 and 1998. Today, this curved hip-roofed powder house sits majestically high on the "great rock," reminding modern-day residents that the past is still very much present in old Dedham town.

DOVER

Dover's large powder house was built in 1800 on a rock on land presented to the town by Captain Samuel Fisher. It measured fifteen feet square and was located on Walpole Street. Ammunition was stored in the building during the War of 1812 and for many years thereafter. It was torn down in 1845.

DRACUT

All that is known about Dracut's powder house was that it was built in 1799 westerly near the house of Henry Richardson, near Hovey Square and Henry Street. It was later used as a well house.

DUXBURY

Other than a vote taken to build a powder house in 1769, nothing else is known about Duxbury's powder house or what happened to it.

EASTON

A powder house formerly stood near Daniel M. Dailey's house in a field that contained a vein of sandstone. It was no longer standing by 1886, when the town history was published and written by William L. Chaffin. No further information could be found.

ESSEX

Essex built a powder house in 1820 at a cost of ninety-five dollars. The building stands atop Fifteen Pine Hill (formerly White's Hill) and was used by the Light Infantry and the local militia for many years. The rectangular building stands roughly seven feet high and five feet wide. It was built of cut granite and probably had a hip roof. The door is located in the center of one of the four walls. The accompanying postcard photo is extremely rare.

Circa 1920 postcard photo of the Essex, Massachusetts 1820 powder house. *Matthew E. Thomas collection.*

FITCHBURG

At a town meeting held on September 8, 1806, Fitchburg voted to form a committee to look into finding a "suitable place for a Powder House, and ascertain what in their opinion will be the best materials for building a powder house; procure a plan for the house, and ascertain the expense as near as they can, and to make a report to the town at some future meeting. Voted and chose Joseph Fox, Asa Perry, & Esquire Israel Turner, a Committee for the above purpose."

At a town meeting held on March 7, 1808, the town voted again to build a powder house and selected Jonas Marshall, J. David Boutwell and Oliver Fox to be a committee to build said powder house.

Fitchburg records fail to mention where the powder house was built or what became of it. There was mention in 1845 of the town voting to build a new powder house, but there was no record of the vote.

FOXBOROUGH

In the early days of Foxborough, the town's powder and lead were kept in the meetinghouse on the town common, but in 1818, a powder house—"eight [feet] square, shingled, plastered and painted"—was built on Powder House Hill, located behind the Foxborough Town Hall. This building stood as late as 1858. When the powder house was dismantled, the materials were used in the construction of the 1858 Reservoir Water Tower, built on the site of the old powder house.

According to a circa 1910 postcard drawing, the Foxborough Powder House was round with a pointed roof.

Circa 1915 postcard illustration of the Foxborough, Massachusetts powder house, which stood between 1818 and 1858. *Matthew E. Thomas collection.*

GROTON

The Groton Powder House was a stone building located on a hillside just south of the meetinghouse near Groton Center. It was built between 1772 and 1774 on land owned by Benjamin Bancroft and originally measured twelve feet square. On July 10, 1777, during the Revolutionary War, it was voted by the State to "enlarge the Magazine [Powder House] at Groton… sufficient to hold five hundred Barrills of Powder." The building was immediately enlarged from twelve square feet to twenty square feet, and within a week, the Massachusetts Board of War deposited five hundred barrels of powder at the magazine in Groton.

This powder house stood on the west side of Main Street (before High Street—now renamed Powder House Road—was built) with only the meetinghouse and horse sheds being its nearest neighbor. Until 1780, the powder house was heavily guarded by the town militiamen, with the requirement that "special care [be taken] that no person be enlisted into the Guard that is not known to be dedicated to the American Cause."

The powder house soon fell into disrepair after the War of 1812, and during the summer of 1829, it was dismantled and the stones reportedly used to shore up a well dug to provide water to the meetinghouse and the barns of the former Hoar's Tavern, later called Stagecoach Inn.

HARVARD

On November 2, 1812, the town of Harvard voted to "erect a house to contain the military stores of the town." It was left to the selectmen to "be a committee to see that the said house is built." The committee decided the building was to be sited "on the rocky hill west of the horse sheds on the Common" and to be built to standard specifications for the time: small, square, hip-roofed, built of brick with a heavy oak door and no windows, set on a solid granite rock foundation. The February 1813 report of the selectmen shows payment of eighty-seven dollars to Phileimon Priest, the builder.

Before the erection of this powder house, the town's supply of ammunition was stored in the roof of the meetinghouse because it was deemed a safe and appropriate storehouse for gunpowder. It was dry and centrally located and had easy access. Risk of fire was minimal, as the meetinghouses of those

This 1812 Harvard powder house is somewhat unusual in that it stands in Harvard Village, close to nearby buildings and a sturdy granite stone-walled town pound used to restrain livestock that escaped from local farms during the eighteenth and nineteenth centuries. *Author's photo.*

days were never heated. At the start of the Revolutionary War, the required amount of ammunition was large enough to warrant a separate building to store it. However, in 1775, the town voted down a warrant article to build a powder house and voted to leave it to the selectmen to find a place to keep the supply of gunpowder. By 1812, the uncertainty of a second war with

Great Britain and the increased bulk of ammunition persuaded the town to approve the building of a powder house.

With the removal of the horse sheds circa 1914, the building became more exposed to the public. Restoration of the building took place in 1914, a new door replaced the old one in 1932 at a cost of forty dollars and in 1948 the building received a new asphalt shingle roof. In 2006, the Harvard Historical Commission continued in this tradition and replaced the old 1948 roof with new sheathing and newly sawn red cedar shingles. The building stands roughly seven and a half feet high and has three granite steps adorned with an ancient boot scraper in front of the powder house door.

A plaque on a large boulder nearby commemorates both the 1812 powder house and the adjacent 1821 town pound. Harvard has every reason to be very proud of the town's longtime devotion to maintaining its historical landmarks.

HARWICH

Harwich built a small square powder house in 1770 opposite Brooks Academy Museum on Route 39 and was used during the Revolutionary War until the Civil War. The original location of the powder house was on Brooks Park but was later moved to Banks Street, near the library, and it was used to store garden tools. Eventually, it became the property of the Larkin family and was given back to the town by the late Anastasia Larkin in memory of her father, Francis P. Larkin. He was credited with recognizing its historic significance and preserving it. The powder house was moved to its present location opposite Brooks Academy in 1975.

The Harwich powder house is probably the smallest such structure in New England, measuring only four feet square by five feet high, with a hip-roof. The building is made of stone and covered with white-painted shingles. This powder house is also unusual in that it has a window in it. The entry door is located on the very left side of the building instead of being centered, as is the case with most New England powder houses.

A fascinating historical anecdote to this historic landmark was written by Douglas Karlson in 2005. It relates to a cannon nicknamed "Long Tom" that was brought to Harwich in 1815 to defend the powder house. In the nineteenth century, residents of Harwich Center and Harwich Port—a section of Harwich—vied for possession of the gun. Occasionally, in the middle of the night, Long Tom was carted down to Harwich Port by

The 1770 Harwich, Massachusetts wooden-shingled powder house is the smallest existing powder house in New England. *Author's photo.*

pranksters and then hauled back the next day. But the cannon also has a tragic history. In 1843, a group of Harwich men decided it would be a good idea to fire off the cannon. It discharged prematurely, badly burning and blinding Harwich resident Cyrus Allen.

Twenty years later, in 1863, to celebrate the Union victory at Gettysburg, Pennsylvania, the cannon was again fired off, injuring several men and blinding one.

One would have thought that a lesson would have been learned after these two terrible accidents. In 1865, a group of Harwich men decided to fire off Long Tom to celebrate the fall of Richmond, Virginia, at the end of the Civil War. According to letters written by Joshua Underwood of Harwich, the cannon was mounted on wooden-spoked wheels and was taken to a field on Parallel Street by Harwich residents Benjamin Franklin Robbins, James Baker, Joshua Robbins and Underwood. As the men loaded the cannon, the ramrod broke in the barrel. They found a crowbar in a nearby barn to finish their preparations. While thus employed, standing in front of the weapon's muzzle, the cannon fired off.

The explosion, probably set off by sparks caused from the crowbar, killed James Baker outright. Baker, twenty-eight, was an abolitionist who had just returned from three years' service in the Civil War. One of the Robbins family members had his hand severed and was carried in agony to the doctor's house, where he died hours later. Underwood received an injury to his arm and ran away from the scene, tearing off his burning clothes. He was extinguished with water by a bystander. To prevent another tragic occurrence, the cannon was spiked, and in 1913, the cannon was mounted in its present location.

Haverhill

In 1767, Haverhill erected the first of three powder houses. The 1767 structure measured eight feet square. Its location remains unknown.

A second powder house was built in 1805 and also measured eight feet square. It stood on the north side of Powder House Lane, now called White Street. It was removed about 1845, when the present powder house was constructed on Golden Hill. This third powder house is a circular brick structure measuring about twenty feet high. It has a tall, conical, shingled roof with pointed brick dentils providing minimal ornamentation under the

Top: Circa 1910 postcard of the 1845 Haverhill, Massachusetts powder house, which stands on Golden Hill and is one of the tallest structures of its kind in New England. *Matthew E. Thomas collection.*

Bottom: In this circa 1860–1870 stereoview of the 1845 Haverhill powder house, it is missing its ornamental weathervane. The adjacent stone wall no longer exists near the structure. *Matthew E. Thomas collection.*

eaves around the building. The roof is capped with a finial that formerly held a weathervane. The door was bricked in at an unknown date.

A fourth powder house formerly stood on Indian Hill in the Bradford section of present-day Haverhill. The dates of construction and eventual destruction are unknown.

Hingham

A red brick powder house built in 1755 stood nearly on the same site as the North Meetinghouse on a hill on Lincoln Street. It was built by the town of Hingham and housed the town's supply of gunpowder for many years. It was later moved in an unknown year to Powder House Hill near Elm Street. Town records fail to mention what became of the building.

Holliston

Holliston originally stored its supply of gunpowder in the meetinghouse. Since there was no fireplace or stove in the meetinghouse, this was deemed a safe place to store it. However, on one sultry Sunday afternoon, a very severe thunderstorm arose, and the congregation decided it would be wise to abandon the building before lightning struck it and ignited the powder therein.

As a result of this fearful event, it was deemed wise to find a safer place to store the town's supply of gunpowder. The Holliston powder house was thus built in 1804. It measured eight feet square and was made of brick and had a brick domed-shaped roof, similar to the Amesbury, Massachusetts powder house. The total cost of construction was sixty dollars. This powder house stood on Powder House Hill on the rocks southwest of John Stone's Tavern in Holliston Center.

The powder house, along with its guns, was sold by the town in 1835. The powder, having been spoiled by the rain from the leaky roof, was thrown out, the building demolished and the materials removed to Medway, Massachusetts, during the summer of 1839 or 1840.

IPSWICH

As early as 1681, a powder magazine was kept in the Ipswich Meetinghouse. In 1702, town records state, "The Selectmen are to have a room in the beams by the clock of the meetinghouse for powder."

In 1792, the town built a circular brick powder house with a conical roof at a place called Gravel Pit Pasture, near Bay Road. It was completed by November 2, 1792, and cost the town £33 old currency. The powder house stood as late as 1871, but unfortunately for this historic early American town incorporated in 1634, famous for its outstanding examples of early American architecture (and fried clams), the town took no further interest in the building, and it fell to ruin and disappeared.

Another powder house was built in 1820 on the hill near the Ezra Perkins house. It disappeared as well. No information as to what type of materials were used in the construction of the building or when it disappeared could be found in the town records.

KINGSTON

In 1806, the town built an eight-foot by eight-foot wooden powder house on the Town Green on Green Street near the burying yard. It originally had a four-sided pointed wooden roof. Local militias would practice their military drills and marches on the Town Green using arms and ammunition stored within the Kingston Powder House.

It is believed by some Kingston locals that the powder house was moved from the Town Green to Evergreen Cemetery in 1924. While there, an addition of about ten feet was added onto the historic structure, complete with two windows and utilized for storage. As of 2013, plans were underway to restore the powder house to its original size and to make it another proud jewel among Kingston's many historic landmarks. It is currently being stored on the grounds of the 1844 Fraunce One-Room Schoolhouse on Elm Street.

The wood-clapboard 1806 powder house in Kingston, Massachusetts, is currently under restoration and seeking a permanent site as of 2013. *Author's photo.*

LYNN

This northeastern Massachusetts town built a powder house at a place called "High Rock" in 1804. The cost of construction was $120.

About the year 1854, an unknown arsonist set fire to the powder house, and it was completely destroyed. However, the storing of powder there had ceased years before its destruction. No additional information can be found in the town records.

LUNENBURG

The Lunenburg powder house was supposedly built in 1818 and was still standing in 1881. It was a square brick structure with a hip roof capped with a small finial. It stood near the Upper Graveyard. It was destroyed by fire circa 1881.

Lunenburg, Massachusetts 1818 brick powder house illustration from *History of Lunenburg, MA 1718–1978*. Published by the Lunenburg Historical Society (1977).

LYNNFIELD

All that is known about the Lynnfield Powder House is that it was built on Powder Hill and that it burned down. Prior to the construction of this powder house, Lynnfield's stock of ammunition was stored in the town meetinghouse.

MANCHESTER-BY-THE-SEA

This town was always called Manchester from its incorporation in 1645 until 1990, when residents adopted the longer name to distinguish it from its much larger namesake city of Manchester, New Hampshire.

On the summit of Powder House Hill, towering above School Street is a square red brick powder house built in 1810. At the time, tensions were mounting between the United States and Great Britain, and the Manchester-by-the-Sea powder house was used during the War of 1812. The powder house continued to be used until about 1850.

The building measures approximately five feet square and eleven feet high. The shingled hip roof is capped with an ornate finial. Because it was hidden by trees for so many years, the powder house has been the target of brutal vandalism over the years. There was evidence that both the inside and outside of the building had been used by vandals as a paintball target. In 2006, the town and local historical society undertook Phase I of the restoration of the building and surrounding property. It was also planned to make Powder House Hill a local destination for residents by pruning back the trees, exposing the powder house and reopening the beautiful sweeping

The 1810 powder house overlooking Manchester-by-the-Sea. *Matthew E. Thomas collection.*

view of the town and harbor from atop the hill. Manchester-by-the-Sea is to be congratulated on undertaking this restoration project and preserving this rare surviving example of New England architecture.

Marblehead

The Marblehead powder house is one of the better-known structures of its type in New England. Its distinctive tapered conical roof makes it a favorite destination for history buffs taking in Marblehead's many historical landmarks and enjoying its illustrious early American heritage.

Marblehead has the distinction of having the oldest powder house still standing in New England that was built specifically as a powder house. It was built in 1755. The oldest existing powder house in New England is located in nearby Somerville, Massachusetts. It was originally built as a gristmill in 1703–04 and then later converted into a powder house in 1747.

The Marblehead powder house was built in 1755 at the intersection of Green and Lincoln Streets at the outbreak of the French and Indian War. Green Street was originally called Ferry Road when the powder house was built. Colonel Jeremiah Lee, Major Richard Reed and Colonel Jacob Fowle were appointed as a committee to build a circular brick powder house. The bulging conical shingled roof is unique in its design among New England powder houses. During the eighteenth century, a "keeper of the Powder House" was appointed annually by the town. In 1802, the town spent fifty cents to paint the ball atop the powder house roof. The structure housed both muskets and powder for the town of Marblehead from the time of the French and Indian War through the Revolutionary War and in the War of 1812. It was most likely utilized for several years after the end of the War of 1812 in 1814.

Time has treated the Marblehead powder house kindly over the past 258 years because Marblehead residents have taken admirable pride in the structure through the years and provided for its upkeep. In 1812, a great deal of repair work was done by the town to the powder house in preparation for the War of 1812, and in 1871, the town voted not to dismantle or sell the powder house, instead voting to restore it and to discontinue storing powder therein. Long may this outstanding relic of early American history continue to adorn the historic streets of Marblehead.

Fort Sewall, at the end of Front Street, was originally called Gale's Head Fort when it was first built in 1644. The fort was rebuilt in 1742, 1775, 1794

Left: A 2012 photo of the historic 1755 Marblehead, Massachusetts powder house on Green Street, one of the oldest existing powder houses in New England. *Author's photo.*

Below: Circa 1905 postcard of the 1755 Marblehead, Massachusetts powder house. This has been one of the most popular tourist attractions in Marblehead for well over 150 years. *Matthew E. Thomas collection.*

The historic 1775 powder house/magazine at Fort Sewall in Marblehead, Massachusetts, was one of many powder houses located in historic old forts around New England. *Matthew E. Thomas collection.*

There are few gun houses still standing in New England. This is the 1810 gun house located in Marblehead, Massachusetts. *Author's photo.*

(with a blockhouse), 1799, 1809 (as an eight-gun fort) and again in 1861 as a twelve-gun battery during the Civil War. It was renamed Fort Sewall in 1742 in honor of Samuel Sewall, who became a Massachusetts Supreme Court justice and was later elected to Congress.

As tensions grew between Great Britain and the American colonies during the 1770s, more money was put into the fort for repair and upgrades. A powder magazine was built circa 1775 behind the existing brick barracks. It was made of bricks and fieldstones and then covered with dirt. The barracks measures roughly twenty-five by thirty-five feet. The barracks have three round arched doors and four similar windows once covered with shutters. Another powder magazine was built in 1794 or 1799. Both powder magazines still exist.

During the Revolutionary War, British ships blocked Marblehead harbor but were never able to progress too far because local volunteer soldiers made the fort look more intimidating than it was. British ships offshore eying the activity in and around the fort with their spyglasses deemed it wise not to attack the fort. Little did they know the fort was out of ammunition!

The United States later gained its independence from Great Britain, and when the War of 1812 came around, additional money was put into fort repairs. On April 3, 1814, two British ships were chasing "Old Ironsides," better known as the USS *Constitution*, off the coast of Marblehead. Seeking protection, Old Ironsides sailed into Marblehead Harbor, where it was protected and defended by the guns of Fort Sewall. Fort Sewall was such an important fort in its heyday that two U.S. presidents visited the fort. In 1817, President James Monroe visited the fort while on a visit to New England, and in 1833, President Andrew Jackson enjoyed dinner in the fort.

In 1863, the fort was enlarged to its present size, and during the Civil War, it housed a few Confederate prisoners of war. Fort Sewall was manned again by soldiers during the Spanish-American War in 1898 and 1899, and this would be the last time it was manned for military purposes.

Ownership of the fort was turned over by the federal government to the town of Marblehead in 1922. Today, it is open to the public, and occasional military encampments take place on the grounds of this historic fort.

In 1808 or 1810 (construction dates conflict), the town of Marblehead built a larger structure to house its ammunition, guns, shot and military supplies. It was a rectangular, two-bay brick structure with large wooden doors. This well-maintained two-hundred-year-old gun house still stands at 45 Elm Street and is one of the few structures of its type still standing in New England.

Medfield

At a Medfield town meeting in 1764, the town voted not to build a powder house. However, in 1768, the small town of Medfield voted to spend ten dollars to build a powder house. It was sited on a large rock atop what has since been called Powder House Hill, just north of Kingsbury Mill Pond. The measurements, construction materials and builders are lost to history.

In 1773, the powder house was struck by lightning. Shortly thereafter, the town voted "to repair the powder house and fix a wire to preserve the powder house from lightning."

The powder house was sold in 1857 for one dollar. It is unknown whether the building burned or was dismantled.

Medford

All that is known about the Medford powder house is that it is mentioned in an article in the September 1, 1774 *Massachusetts Gazette*, which stated, "260 British troops seized 250 half barrels of gunpowder at the Charlestown Powder House (now the Somerville, Massachusetts powder house) on that day, and that the British were then proceeding to the Medford Powder House to seize the powder there." Nothing else is known about this powder house.

Medway

A powder house was built in Medway in 1810 on the farm of Edward Clark to store the town ammunition. No further information can be found on this structure.

Middleton

Prior to the outbreak of the War of 1812, Middleton residents were becoming increasingly alarmed over the diminishing relationship between Great Britain and the United States. Consequently, the town began preparing for the possible outbreak of hostilities.

A town committee was formed in 1811 for the purpose of procuring land for a powder house and seeing that it was properly constructed. The committee consisted of Daniel Fuller, Andrew Merriam and Captain Joseph Peabody. They decided to erect the building on a hill southeast of the Middleton Meeting House (now east of Pinedale Road). This land was John Fuller's pasture. For one dollar, Fuller gave the town a deed to the spot on which the town built a powder house, granting himself the privilege of passing and re-passing through the land. It is believed that John Fuller also constructed the building.

The powder house was built eight feet square and seven feet high, with a wooden door and roof. It was set on a stone foundation. Regretfully, all traces of this old landmark have disappeared.

MILLBURY

In the village of Millbury, on a small knoll located to the southeast of the imposing 1826 Asa Waters Mansion at 123 Elm Street, is an outcropping of rocks with a rectangular granite rock structure built on it. It measures about fourteen feet long by ten feet wide, with thick granite rock walls and no doorway entrance and the roof has long since disappeared. Apparently, no one associated with the local historical society seems to know what the structure was actually used for.

This 2009 photo shows remnants of what is believed to have been the Millbury, Massachusetts powder house. *Author's photo.*

Therefore, to be on the safe side, this author deemed the structure curious enough to warrant placing it within this book. Hopefully, after some in-depth research in the future, someone might be able to solve the mystery and reveal that the existing remains are possibly those of an old granite powder house.

Milton

The 1811 Milton, Massachusetts powder house is very typical in architectural style to numerous powder houses located around New England. *Author's photo.*

The Milton powder house is a small brick structure, measuring eight feet square and eight feet high. It has a two-and-a-half-inch-thick wooden door and a shingled hip roof. It is believed to have been built in 1811 by Benjamin Bronsdon on the corner of Poor House Lane and Canton Avenue. It cost $100 to construct.

During the War of 1812, part of the Milton selectmen's duty was to meet at the powder house and fill cartridges for the town musters.

To preserve this historic structure, the Milton Historical Society moved the powder house in October 1974. It is now located across the street from the Milton Town Office Building on Canton Avenue.

NEEDHAM

According to the Needham town history compiled by George Kuhn Clarke in 1911:

> *On July 12, 1753, Lieut. Amos Fuller, Timothy Newell and Lieut. Aaron Smith were chosen to procure a place to Keep the Town Stock of Arms and Ammunition. This powder-house cost 6 pounds, 8 shillings 2 farthings but it does not appear whether it was in the meeting house or was an independent building.*
>
> *When the meeting-house burned in 1773 town ammunition was kept in the loft, apparently in a chest, and the first distinct powder house stood on Powder Point Ledge, located west of Nathaniel Wale's barn.*

In 1764, Michael Metcalf was paid "four pence for a Staple for ye powder house," and on June 2, 1777, the town voted "to Sell ye powder house and chose Col. McIntosh to Vendue [auction off] the powder house."

In 1796, the town refused to build another powder house, but in 1800, Colonel Kingsbery, Daniel Smith, Captain Ebenezer McIntosh, Colonel Alden and James Smith were chosen as a committee to build one. The 1800 powder house cost upward of $100.00. James Smith and Daniel Sanger were the masons who laid the bricks. There is a payment of $7.50 for 1,000 bricks delivered and also one of $2.00 for carting a like quantity from Watertown.

In 1835, Israel Whitney took away the powder house. Traces of it remained on Powder House Ledge as late as 1911.

NEWBURY

As early as May 1743, the selectmen of Newbury were authorized to build a convenient house "in which to keep the town's stock of ammunition," but they evidently failed to act on the authorized town vote. They were again authorized in October 1745 to build a powder house, and still nothing was done.

At another town meeting held March 12, 1753 or '54, the selectmen were ordered to build a powder house. This powder house was built next to the Old Hill Burying Ground in present-day Newburyport, which, until 1764, was a part of Newbury.

On February 19, 1754 or '55, the selectmen paid Abiel Somerby £6.5.4 old currency "for bricks for ye Powder House," and a few weeks later, they paid several small bills for shingles, rocks and labor on the building.

It is not known when this powder house disappeared, but it was still in existence in 1771 because it is shown on a partial map of Newburyport made that year by John Vinalfeat.

NEWBURYPORT

Newburyport was originally a part of Newbury until it separated in 1764. Therefore, the powder house of 1754–55 has been described previously under Newbury because the site was actually in present-day Newburyport. It stood near the present-day Bartlett Mall on the side of the Old Hill Burying Ground. Many have heard of the Boston Tea Party in December 1773, but few are aware of similar tea parties that also took place around New England. Early in 1775, Newburyport had its own "tea party" when several chests of tea were seized from the 1754–55 powder house and destroyed after being placed within it for safekeeping.

Newburyport records show the following individuals held the position of powder-house keepers. In 1767, John Downer was selected to the position on June 27, but John Bromfield was listed as the new keeper less than two months later on August 10. On April 4, 1771, the selectmen appointed Paul Shackford as keeper, and he probably held the position until March 12, 1779, when John Little was appointed. Mr. Little was instructed "to receive in and deliver out Powder & Keep a record of the same & not to Suffer [allow] any Person to enter said [powder] house with Shoes on." This was because at that time, shoes were made of nails, and if someone stepped on a stone creating a spark in the powder house, the possibility of an explosion could have devastating consequences.

About the year 1794, a stone powder house was built by Daniel Pierce on his farm. Tradition relates that on one occasion, a slave owned by Mr. Pierce placed a lighted candle in a keg of powder. Eventually a fire started, and the powder ignited. The resulting explosion blew out one side of the house. A slave who lodged there was blown, bed and all, into the limbs of a large apple tree, very much to her great amazement.

On November 28, 1820, Newburyport selectmen were asked to choose a safe and convenient location for a new powder house. A town vote on April

Godfrey's Hill powder house in Newburyport, Massachusetts, was restored in 2011. *Author's photo.*

11, 1822, "accepted the selectmen's report which recommended that a new powder house be built on Godfrey's Hill near Common pasture on Low Street as the most suitable place on which to erect a Powder House." This large circular brick structure with a conical shingled roof capped with a small finial still proudly stands. It cost $494.65 to build, according to Newburyport town records.

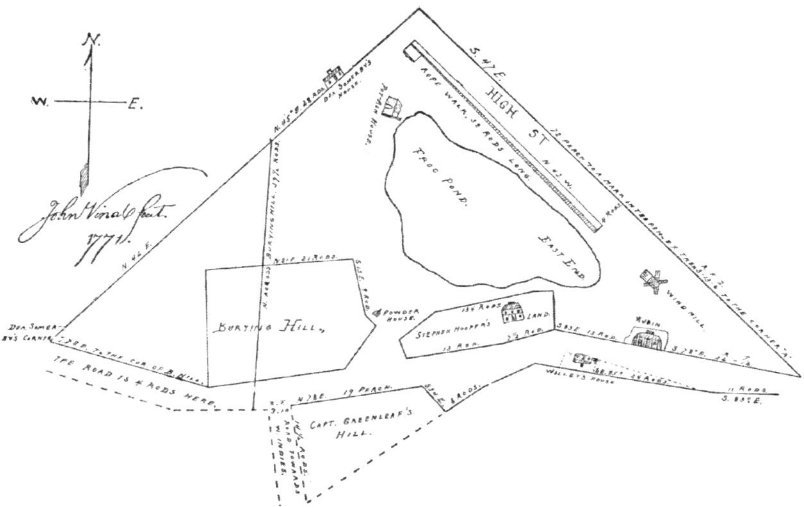

This 1771 map of part of Newburyport, Massachusetts, shows the location of the original powder house formerly located beside the historic old Newburyport graveyard near Bartlett Mall. This graveyard is the burial place of "Lord Timothy Dexter b.1737–d.1806," who lived in Newburyport and became one of New England's most eccentric and colorful characters. He briefly lived in both Poplin (now Fremont) and Chester, New Hampshire, during the late eighteenth century, where he conducted some of his eccentricities. *Courtesy of New England Historical Research Associates.*

Late nineteenth-century photo of Newburyport powder house on Godfrey's Hill. *Photo courtesy of Newburyport Archival Center, located in Newburyport Public Library.*

This historic structure in a city renowned worldwide for numerous architectural landmarks was partially restored in 2009 after six years of trying to acquire permission from the Massachusetts State Armory and National Guard (on whose land the powder house has rested since 1950, when the city deeded to the state the land for an armory erected in 1957). In 2009, the State of Massachusetts deeded a ninety-nine-year lease for a nominal fee back to the city of Newburyport containing the powder house and .4 acres. The nonprofit Newburyport Preservation Trust and the Powder House Restoration Committee will oversee and fund the restoration and future operation of the building.

The powder house had been subjected to vandalism and deterioration since 1976, when it was last repaired. Serious roof leaks, falling brick walls and rotting interior beams all need repair or replacing. When complete, the restoration will cost approximately $200,000. The city and Preservation Trust are to be congratulated for their perseverance in seeing this worthy project through.

NEWTON

Newton constructed its powder house in 1799. The brick building stood between Lyman Street and the Town Common.

There is no information available that indicates the size of the building, what materials were used, how much it cost or when the powder house was destroyed. All that is known is that the building and its contents were sold in 1869.

NORTON

For many years, the town's stock of powder and ammunition was stored in the garret of the old meetinghouse. However, many residents were of the opinion that this was an unsuitable place for military stores.

On April 2, 1804, the town voted not to build a powder house but had a change of heart at the June 12, 1805 town meeting. It was voted to build a powder house of brick and lime and to auction off the project to the lowest bidder. The winning bidder is not known, but the small circular building measured six feet in diameter and six feet high. It had a conical wooden roof. It stood behind the Trinitarian Meeting House on East Main Street. It is not known when the building was taken down.

According to George Clark's *History of Norton*, published in 1859, "The powder house contained one old rusty flintlock musket and bayonet, two kegs of ball cartridges so completely saturated with water and petrified that it is doubtful whether they would burn in any common fire, a half-bushel of bullets, a peck of flints, three old tin pails, a few broken boxes, and almost hoopless powder kegs."

Unlike many town historians who lamented the loss of their town powder houses, Clark wished the Norton powder house to "peacefully go to ruins, with none to weep over its moldering remains."

Oxford

The Town of Oxford built its powder house after a favorable vote at the town meeting in 1806. The land for the building was leased to the town by Mr. Sylvanus Town. The seven-foot-square brick structure with hip roof was built near the north end of the town common. The powder house never received much upkeep by the town. After a vote in May 1837, the selectmen were authorized to dispose of both the powder house and its contents. It was moved to a nearby farm, where it was used for many years as a smokehouse. A descendant of this farm family, Leslie J. Chaffee, presented the former powder house to Old Sturbridge Village early in 1957 to be used as part of an outdoor historical museum complex. The Oxford Powder House can now be seen sitting atop a small hill in Old Sturbridge Village behind the Town Pound.

Peabody

The gun house stood near the town common in Peabody and was described as a small wooden structure, built of vertical siding and covered with a slate roof. It was just large enough to accommodate an artillery piece and was later used to house a town fire engine during the late nineteenth century.

Petersham

The Petersham town history notes that this small rural community's powder house was built in 1817. It was six feet by seven feet and stood six and a half feet high. It was one of very few powder houses built of wood and was estimated to have cost fifteen dollars. When the town history was published in 1948, the powder house was believed to have stood on the north side of East Street, between Albert Brunelle's house and the Jackson House. Nothing remains of the structure.

Pittsfield

The scenic western Massachusetts community of Pittsfield had constructed a good-sized powder magazine, but the actual date of construction remains unknown. Disregarding public safety, the building was located in the northeast corner of the cemetery, near the center of town.

Even though there are no physical descriptions of this powder magazine, it had a colorful history. According to the *1876 History of Pittsfield* by J.E.A. Smith, the following powder magazine explosion occurred in 1838.

> *At the time of the explosion, the powder magazine contained about 700 pounds of gunpowder; of which 400 pounds intended for use in the testing of arms manufactured at Lemuel Pomeroy's armory, was owned by the United States government, and the remainder by various merchants.*
>
> *For several months prior to the explosion the residents of Pittsfield Village had been annoyed by the night-time mischief of unruly young men. Several petty crimes such as removing signs, defacing fences, and unhinging gates were endured with little resistance by those who suffered from them. Finally, the artillery-pieces belonging to the state were frequently taken out of their house and discharged in the dead of night, while the firing of muskets and smaller arms was incessant night after night. Emboldened by the neglect of efficient measures to bring them to justice, the troublemakers, sometime previous to the Fourth of July, stole both the cannon and hid them until, on the night of the seventh of July, one of them was planted in the earth in front of the house of Lemuel Pomeroy, and after being loaded to the muzzle was discharged, ruining the piece and inflicting considerable injury upon the house of Mr. Pomeroy and to those opposite it.*

Proceeding to more dangerous practices, the disturbers of the peace entered the powder house several times, by means of false keys, and obtained considerable quantities of gunpowder.

Now, at last, measures were taken to bring the culprits to justice; and, at the request of Mr. Pomeroy, the Adjutant-General ordered the artillery pieces to Boston. Provoked by this threatened interruption of their dangerous exploits, the culprits threw out menaces to which little attention was paid, that being the (powder) magazine should be exploded. Even their previous recklessness had not prepared local authorities or the people to believe that they would perpetrate an act fraught with such danger to life and property. No watch was therefore set over the exposed property, or any other precautionary measures taken. On the night of July 12th, the threats were put into action, and at 11:30p.m. the powder magazine was exploded, the great mass of gunpowder contained in it having been fired by a slow-match.

By remarkable good fortune, no loss of life ensued; but the destruction of property was extensive. The house owned by Nelson Strong, situated on Fenn Street, very near the magazine, was a perfect wreck. Windows, doors, and whatever else could be loosened by the explosion were thrown upon the beds where the family was sleeping. Every piece of crockery in the house was broken. On the opposite side of Fenn Street, the house of Mr. Henry Callender was severely injured in a similar way, and a brick thrown entirely through it. The brick school-house on the corner of Pearl & Fenn Streets, and a neighboring house lost their windows. The dwelling house of James Warriner on East Street, James H. Dunham on North Street, the Medical Institution, the town hall, the Congregational and Baptist Churches suffered severe damage. Some twenty other buildings suffered considerably, but chiefly from the breakage of glass.

Intense excitement prevailed on the following morning, and at a large meeting of the citizens, after speeches such as the occasion would naturally call out, a committee was appointed to bring the perpetrators to justice. The selectmen offered a reward of \$200.00 for their conviction. The committee reported on July 19th that they were "engaged in a course of investigations and suits which promised to bring the guilty parties to justice."

Nevertheless, several young men connected with highly respectable families were generally believed to be implicated, no one was ever convicted.

PLYMOUTH

At a Plymouth town meeting held May 16, 1757:

> *The Committee chosen by the Town of Plymouth at their meeting in March last, to take into consideration the safety of the Town Stock and where & how it may be best to keep the same, have considered this affair and are of the opinion that the Almshouse* [poor house] *where the town's stock is now kept is not a Place of Safety* [for the powder] *nor can they recommend any other Place already built as more Safe, and that unless the Town shall cause to build a small Powder House in some convenient Place the said stock is as safe where it now is* [the Almshouse] *as it can be in any other place.*

Apparently, the town decided not to build its first powder house as early as 1757.

Plymouth did form a powder house committee in early 1770. Thomas Foster, Esquire; John Torrey; and Elkanah Watson were to procure a suitable place for building a powder house and to provide a cost estimate for the building.

The committee gave its recommendations at the March 26, 1770 Plymouth town meeting. It was proposed to build a powder house "on the level spot on the hill about a rod to the westward of Mr. Robbins fence that encloses the parsonage being the northerly part of Fort Hill." The committee had two estimates for consideration. If the town chose a thirteen-foot diameter, six-foot-high powder house, the cost would be £25 old currency. If the voters preferred a twelve-foot by six-and-a-half-foot-high structure, it would cost £20 old currency. The town approved the committee's recommendations and selected the £20 option.

The round powder house had walls a brick and a half thick and a brick roof. The builders were Deacon Torrey, Captain Elkanah Watson and Isaac Lothrop. Pictures of the original brick powder house show that it was encased in a shell of wood construction, octagonal in shape, at some unknown later date. This 1770 brick powder house disappeared circa 1880. Fortunately, photos of the original structure encased in wood survived and were made into postcards at the turn of the twentieth century.

At a Plymouth town meeting held May 18, 1772, townspeople voted to enact a town law requiring "all inhabitants, traders, and importers living or visiting within the town, to bring any large or small quantity of gunpowder which they are possessed of, to the powder house for safekeeping within 24 hours of entering the town."

Left: Circa 1910 postcard of the original eight-sided, wood-shingled Plymouth, Massachusetts powder house that formerly stood in the northwest corner of Burial Hill Graveyard. It was replaced in 1920 with a new similar-looking brick powder house near the original site of the 1770 structure. *Matthew E. Thomas collection.*

Below: The 1920 brick replica of the 1770 Plymouth powder house was dedicated during an elaborate ceremony in 1920 attended by hundreds of spectators. *Author's photo.*

Failure to do so could result in a fine of twenty shillings for each and every such offense if found guilty in a court of law. The town treasurer was responsible for prosecuting persons who did not comply with this law: "he shall pay the sum of 20 shillings for every neglect, to be recovered by a lawsuit brought against him by the selectmen of said town. All fines will go to support the Poor of Plymouth."

Upon submitting his quantity of gunpowder to the powder house keeper, the owner had to pay the keeper

> *12 pence per barrel, 6 pence per half barrel, for the first month of storage, and four pence per barrel and two pence per half barrel for every month after the first month. Out of this fee will come the keeper's fee for taking care of the powder house and powder, and if in the Judgment of the Selectmen, the money received for storage is not sufficient reward for the Keeper's trouble, they are to make him further allowance, as they think reasonable, which is to be paid out of the town treasury.*
>
> *The selectmen of Plymouth for the time-being will appoint the keeper and also give orders from time to time for regulating the powder house and preserving the powder put in the building. The Keepers of the powder house will receive and secure powder and ammunition, keep an exact account of all the money received for storage, and render their accounts to the selectmen once a year, or more if required.*

The Plymouth powder house was the only known structure in New England that was used to generate income for the town. It was certainly the most regulated.

The town built what was claimed to be an exact replica of the original 1770 structure on the same site in 1920, but the roof is shaped a little differently. This is the newest powder house standing in New England. A grand dedication ceremony took place in October 1920 attended by a large crowd. The new powder house was draped with a large American flag during the celebration. This landmark still stands on Burial Hill as of 2013.

QUINCY

A powder house once stood near Scotch Pond in the North Precinct section of Braintree, which was later separated and incorporated in 1792 as Quincy.

Prior to the outbreak of the Revolutionary War, many residents of the Middle and South Precincts of old Braintree believed that the North Precinct (now Quincy) sympathized with Great Britain. In 1774, the North Precinct was where the Tories (who supported the British during the Revolutionary War) lived.

In September 1774, when the British were trying to secure all the provincial ammunition in the colony of Massachusetts (like they did at the powder house in Charlestown—now Somerville—on September 1, 1774), an English soldier was seen lurking around the north commons where the Braintree (now Quincy) powder house was located. Upon receiving such alarming news, a force of about two hundred colonists went to the powder house to secure the ammunition. They succeeded in doing so easily.

There is no information that describes the size, shape, cost or builders of this pre–Revolutionary War era powder house or when it disappeared. All that is known is that it is not mentioned as being in existence when the *History of Old Braintree* was published in 1878.

Rehoboth

All that is known about the Rehoboth powder house is that it formerly stood on the southwest corner of the Village Cemetery near Pond Street. No further information could be found.

Rowley

The Town of Rowley originally stored its gunpowder in the spire of the town meetinghouse. In August 1777, the meetinghouse was struck by lightning, but fortunately the powder did not ignite.

The town eventually decided to construct a powder house in 1794 on the rise of ground northwest of Fairview. It was built of brick and lime, was circular in shape, and was of the bigness of a hut. It cost $62.42. In 1848, the town voted to sell the powder house at auction. When a history of Rowley was published in 1946, only a few bricks marked the spot where the building once stood.

ROXBURY

A powder house once stood on Magazine Street on Pine Island, a part of the old Governor Bernard Estate. It was still standing in 1832. Unfortunately, nothing else is known about it.

An interesting account tells of the Massachusetts Colony's stock of gunpowder exploding in a house at Roxbury in 1645. John Johnson, the surveyor general of arms and ammunition for the Colony of Massachusetts, lived in a fair house in the midst of town (Roxbury) when in February 1645, according to Governor John Winthrop's journal:

> *Johnson's house caught fire during the day time, no man knowing what the cause was, and there being in the house 17 barrels of the country's powder, and many arms, all was suddenly burnt and blown up, to the value of 4 or 500 pounds, wherein a special providence of God appeared, for, he, being* [away] *from home, the people came together to help and many were in the house, no man thinking of the powder till one of the company reminded them of it, whereupon all withdrew from the house, and soon after the powder took fire and blew up all about it, and shook the houses in* [nearby] *Boston and Cambridge, so that men thought it had been an earthquake, and carried great pieces of timber a great way off, and some rags and such light things beyond Boston meeting house, there being a stiff gale south, it drove the fire from other houses in the town (for this was the most northerly) otherwise it had endangered the greatest part of the town.*

Roxbury was annexed to the City of Boston in 1868.

SALEM

Salem has the distinction of having the earliest known powder house in New England. The first of three powder houses was built by Samuel Archer and William Allin and was completed at the end of August 1637.

A second powder house was built by the town in 1776. It stood at the west end of Salem.

In 1794, a third powder house was built in "Great Pastures," a section of Salem. It was still standing in 1849 but has since disappeared. Materials and styles of all three of Salem's powder houses are not recorded.

Salisbury

The Salisbury powder house was built in 1810. It still stands on Powder House Hill (formerly Brown's Hill) on land annexed to the adjoining Town of Amesbury in 1888. See Amesbury for more details.

Sherborn

The quaint town of Sherborn built its first and only powder house in 1800 to provide a safer place to store its supply of gunpowder, which had been kept for many years in the town meetinghouse. The move from the meetinghouse was much to the relief of local womenfolk who, presumably, were quite fearful of a potential lightning strike to the meetinghouse during Sunday sermons. The structure was a round brick powder house capped with a conical shingled roof. It stood at the crest of Brush Hill off present-day Powder House Lane. James Bullard was the first powder house keeper. Later, Captain Daniel Paul (1787–1855), a most prominent public official of Sherborn, also served at one time as the "Keeper of the Powder House." The Town of Sherborn voted to demolish the powder house in 1857.

Somerville

This is the most famous, largest and oldest existing powder house in New England. The colorful history associated with this imposing Somerville structure also makes it one of the most extraordinary Revolutionary War–era landmarks still standing in the United States.

The Somerville Powder House, located in Nathan Tufts Park, is the primary example of why these nearly forgotten old powder houses dotting the New England landscape have become such vitally important historic landmarks. It was from several of these powder houses that the opposing enemy strived to capture all-important gunpowder, arms and ammunition stored within these structures to help further their own military objectives. These brick, granite or wooden landmarks became true arsenals of American independence as they quietly sat atop small remote hilltops or on the outskirts of town commons year after year waiting for the next

Rare circa 1860 stereo view of the front side of the historic 1703–04 Somerville, Massachusetts powder house, the site of the first and only successful British capture of Patriot gunpowder in New England on September 1, 1774. *Matthew E. Thomas collection.*

town muster or next war, when they would be useful again. Add to this the fact that they also stored dangerous explosives on a daily basis that could wreak unbelievable havoc on any community if the powder house wasn't managed and accessed in a safe and cautious manner. Better to have highly dangerous explosives stored in a secured town powder house than have them explode in one's house or barn, accidently killing or injuring loved ones or innocent bystanders.

Somerville was originally part of historic Charlestown, Massachusetts, until it separated and was incorporated as its own town in 1842. Somerville

A 2013 photo of the impressive 1703–04 Somerville powder house. It stands as a proud reminder of struggling days some 238 years ago when New England colonists bravely fought to break the shackles of British tyranny and win American independence from King George III of Great Britain between 1775 and 1783. *Author's photo.*

Circa 1874–75 stereoview of the rear side of the 1703–04 Somerville powder house. This structure was first used as a windmill and converted into the Massachusetts Provincial Powder House in 1747. This stereoview was taken in time to celebrate the 100th anniversary of the epoch 1775 Battles of Lexington and Concord. *Matthew E. Thomas collection.*

today is a vibrant community that quickly grew during the nineteenth century and became a city just thirty years later, in 1872. The city is currently a culturally diverse community consisting of over seventy-seven thousand residents located three miles from Boston.

This brick and stone circular structure built in 1703 or 1704 is cone-shaped and stands 30 feet high with 2.5-foot -thick walls measuring 60.75 feet in circumference on the outside. The powder house started its existence as a grist- and windmill on "Quarry Hill" and formerly had three lofts with thick oak beams and flooring, but these were removed in the 1880s, exposing the whole interior from dirt floor to roof. From the door on the north side of the powder house to the opposite side of the building, the interior measures 14 feet and 2 inches. The tapering conical roof is shingled, and there is but one window in the whole building. During much of the nineteenth and twentieth centuries, a 20-foot-tall flagpole adorned the top of the powder house.

This ancient structure is believed to have been built by Jean Mallet, presumably a French shipwright, who came to America circa 1685. He later became a miller and purchased the property in 1703–04, but no evidence exists that definitively verifies he was the actual builder. Jean Mallet died about 1720 and left the old gristmill to his son Michael, who sold it in 1747 to the State of Massachusetts to be used as a powder magazine. The millstones had probably ceased grinding grains many years before the sale.

In 1890, the heirs of Nathan Tufts deeded to the City of Somerville about sixty-five thousand square feet of land, including the powder house. In 1891, additional land was purchased next to the powder house property for the purpose of turning the entire enlarged site into a public park.

In its early days, the tall round powder house stood like a lone sentinel atop Quarry Hill, diligently keeping watch over the former fields and highways of old Somerville. Around it are clustered legends, curiosities, historic events and other charms of antiquity that make this one of the most interesting early American architectural curiosities in New England.

Despite all this varied history, one event in particular catapulted this landmark into the American history books. At the outbreak of the Revolutionary War, this powder house contained the largest supply of gunpowder in Massachusetts. Stored within its massive stone walls were 250 half-barrels of gunpowder. On September 1, 1774, British general Gage sent 260 British soldiers up the Mystic River in thirteen boats. The expedition landed at a place called "Ten Hills Farm" and then proceeded to the Charlestown (Somerville) powder house. Here they seized all the barrels of powder therein and transported them to the castle (fort) in Boston Harbor for safekeeping from the Americans. It was part of a British attempt to confiscate whatever ammunition was in area powder houses to keep military stores from falling into the hands of American Patriots. The capture of gunpowder by the British greatly alarmed American Patriots and sympathizers, and soon rumors were spreading throughout the Massachusetts countryside. This event had been the first hostile demonstration made by British troops, and it caused hundreds of men to come from all over the

This 1930 Massachusetts State Historic Marker details the history of the Somerville powder house/magazine, designating it one of the most historic and significant Revolutionary War era landmarks still standing in New England. *Author's photo.*

region on September 2 to form militia groups that were prepared to fight the British if need be. This event was known as the "Powder Alarm." The tension was eventually diffused, and the men went back to their hometowns. This unexpected hostile response from so many Americans caused the British to call for reinforcements from London. For the Americans, the confiscated gunpowder from the powder house by the British impressed on American revolutionary leaders the urgent need to become better prepared and to form Committees of Safety in various towns throughout New England to expedite better communication and military responses in the event of future British military actions against patriotic American colonists.

During the Revolutionary War, this powder house was in the hands of the Americans and became the facility used by revolutionary forces to house military stores that were eventually used to help support American militiamen besieging the British in Boston and Dorchester Heights, outside of Boston.

Southborough

In 1735, Timothy Brigham proposed and got approval for the construction of a powder house within the stone walls of the town burial ground. It was to house the town's supply of gunpowder, ball and shot. The burial ground was becoming crowded by 1840, so the town voted to empty the powder house of its contents and took it down.

Springfield

Although Springfield was only a small, struggling village during the mid-eighteenth century, its location at the intersection of major highways and the Connecticut River—yet far enough upstream to be safe from enemy attack—made it an ideal place to build an armory/arsenal. Supplies, skilled manpower and adequate waterpower for manufacturing small arms and ammunition were all close at hand. Colonel Henry Knox, General George Washington's chief of artillery, concluded that "the plain just above Springfield is one of the most suitable spots" for the location of an arsenal, and in 1777, the arsenal at Springfield was constructed to manufacture cartridges and gun carriages for the American Revolution. During the Revolution, the arsenal stored muskets, cannon and other weapons.

A powder magazine was built in Springfield in 1778 as part of the arsenal/armory, which soon became a very busy site with workmen of various trades engaged in the manufacture of ammunition, guns and wagons. The powder magazine was filled from top to bottom much of the time and stood in a hollow near the northerly part of the Training Field on land near to where the barracks and laboratories were built. The 1778 powder magazine had disappeared by 1804.

In 1782, a second powder magazine was erected not far from the Boston Road at the easternmost point of the training field located on a hill on the outskirts of town. This long brick powder magazine stood until 1842, when it was taken down because it posed a danger to nearby residences.

Shortly after the end of the American Revolution, during the mid-1780s, an epoch era of depreciating currency and high inflation wreaked financial havoc throughout the former thirteen colonies. As a result, countless citizens found themselves debtors and were jailed for non-payment of various debts. This widespread consequence infuriated many, especially in the poorer classes, who thought that after winning the hard-fought American Revolution, a new dawn of freedom, justice and economic stability would emerge. Instead, the war caused high inflation and severe suffering for thousands in the newly formed nation.

This economic pain caused disgruntled veterans of the Revolution and poverty-pinched farmers of the hill country towns of western Massachusetts to form a mob of 1,500. This mob, led by the now infamous Daniel Shays of Pelham, a former captain in the Revolutionary War, attacked the Springfield Arsenal in the dead of winter on January 25, 1787. Their goal was to capture the various military supplies from the powder magazine within the arsenal and utilize them to force the closure of the state and county courts that were taking their lands for debt payment. Confronted by the cannons and artillery of an organized state militia numbering 1,200 under the leadership of William Shepard, the mob was repulsed. Four of Shays's men were killed and twenty wounded. The overall effect of the volley was the disruption of the rebel morale, and despite other smaller pockets of rebel conflict throughout the region in the months that followed the January attack, the consequent proof in the face of grim resolution was that Shays's cause gradually became a hopeless one.

Derwent S. Whittlesey, author of *The Springfield Amory*, states in his published dissertation:

> *The citizens of Massachusetts as well as those in other states learned two practical lessons from Shays's failed rebellion. First, that the mockery of government under the Articles of Confederation that preceded the adoption of the United States Constitution in 1788, could no longer be tolerated since that political entity had failed to protect its own property, and to provide a sound financial system which alone would avert depreciated currency and approaching bankruptcy. Second, that armed revolt in a democratic republic was untenable, and could lead only to domestic woe and political annihilation.*

As a result of Shays's attack on the Springfield arsenal powder magazine, another valuable lesson was learned. Thus the Continental Congress, which sat in Philadelphia, Pennsylvania, during the summer of 1787, only a few months after Shays's attack on the Springfield Arsenal, encouraged the people of Massachusetts to ratify the adoption of the United States Constitution on February 6, 1788, in part to avoid repetition of another such unfortunate and dangerous revolt. The United States Constitution became the law of the land on June 21, 1788, when New Hampshire became the ninth and deciding state out of thirteen to make it official. Interestingly, a Paper

Money Riot had also occurred at Exeter, New Hampshire, in September 1786, when two hundred rebels surrounded the Exeter courthouse and were quickly dispersed by a hastily organized local militia.

In 1794, the Springfield arsenal/armory was selected by President George Washington to be one of two sites for the establishment of a federal arsenal. It operated until 1968, when the facility was closed and designated as a National Historic Park.

Sterling

All that is known about Sterling's powder house is that it stood near the Congregational Church. It was in existence as early as 1826 because it was mentioned in an article published in the *Worcester Magazine & Historical Journal* for that year.

Stow

A brick powder house once stood on Pilot Grove Hill above Hillside Cemetery. It was there in 1775 but was removed in 1849 for the purpose of enlarging the cemetery. No further information could be found.

Sudbury

In 1771, Sudbury voted to build a powder house in which to keep the town's stock of ammunition. It appropriated "7 pounds 9 shillings and 4 pence, and agreed with Col. John Noyes to build it and place it near, or on William Baldwin's land near Major Curtis," according to the town history.

Another record for the same year states that "the town voted to erect the powder house on the training field near Mr. Elisha Wheeler's." In 1773, the town voted to "remove the powder house to some suitable place on or near the gravel pit hill, and choose a committee to remove the same, if the committee should think the powder house will be sufficient for the use it was built for, and rough cost and underpin the said building."

The powder house was still standing in 1830 because town residents were given permission to erect new tombs on the east side of the powder house.

In 1777, according to the following court directive, the Sudbury powder house was to be utilized as a Massachusetts government powder house, and a small group of soldiers were selected to guard the powder house during the remainder of the Revolutionary War. It stood near a gravel pit on the west side of the river.

> *September 2, 1777.*
>
> *Agreeable to an order of court directing me to enlist 1 corporal and six privates to serve as guards to the stores (of gunpowder & ammunition) belonging to the state deposited at Sudbury, I have enlisted…*
> *Corporal Robert Eames*
> *Silas Goodnow, Jr.*
> *Philemon Brown*
> *Elisha Harrington*
> *Jonathan Graves*
> *Lemuel Goodnow*
> *Jona. Clark*
> *All of Sudbury, & returned the enlistment to Capt. Isaac Wood, Commander of the Continental Guard in said Town.*
> *—Thomas Plympton.*

SUTTON

At a town meeting held on June 22, 1801, voters in Sutton approved the construction of a powder house to put the town's stock in. No other information is available.

TAUNTON

Taunton's powder house was built at an unknown date. It stood on Powder House Hill on the east side of the river, between Plain and Berkley Streets, east of Beacon Street.

After the Revolutionary War, a Major Seaver, brigade inspector, came to Taunton to inspect the powder house. This is the first reference to the building in town records.

In 1799, a record of the powder house contents listed "400 pounds of powder, 750 pounds of shot, and 600 flints." In 1827, the contents included "692 pounds of balls, 18 to the pound, 5 pails, 22 iron kettles, 13 tin basins, 12 tin cups, 7 earthen plates, 9 guns, 9 gun locks, 12 wormers, 1 axe."

The decline of the local militia between 1830 and 1840 caused the powder house to suffer neglect. By 1834, its contents were entirely worthless. The building continued to decay and was eventually taken down. It was the only powder house built in Taunton.

Tewksbury

Though the Tewksbury, Massachusetts and Bedford, New Hampshire Powder Magazines are not among the typical powder houses/magazines generally referred to within this book, they still warrant inclusion because they represent the huge potential for disaster that any New England powder houses could visit upon their communities. Both Tewksbury and Bedford powder magazines were used to store newly manufactured powder and dynamite as part of a large gunpowder and/or dynamite manufacturing facility. The powder magazines in both Tewksbury and Bedford blew up and brought terrible tragedy and chaos to their respective communities. The Tewksbury facility included two powder magazines that were part of the U.S. Cartridge Company and located underground off Billerica and Acton Streets in a section of Tewksbury called Wigginville and Riverside Park that was later annexed to the City of Lowell in 1905.

On July 29, 1903, at 9:00 a.m., these two powder magazines exploded in a spectacular blast that killed twenty-two people and injured sixty to seventy others. All fatalities took place in Tewksbury.

Within a half mile of the explosion, 307 houses and buildings were impacted (125 of them seriously) by the shock of the explosion. Many people were killed or injured from flying bricks and collapsed buildings, as well as the shockwave that killed two children instantly while they were swimming nearby the two powder magazines. Many bodies were terribly mangled and thrown violently through the air. One woman was thrown twenty-five feet and killed. The wounded were screaming and running around hysterically. Five minutes after the explosion, the scene was one of gruesome and awful horror.

According to the *Lowell Sun,* six-year-old Alfred Catherwood, of 48 Epping Street in Wigginville, was playing in the yard with a playmate when the "big blow" occurred.

> *"All of a sudden there were bricks flying all around," he remembered. A fellow who lived five or six houses away from Catherwood's home got hit with a brick instantly killing him. Catherwood later recalled that he "had never seen anything like it, not even the 1975 gas explosion on Bridge Street in Lowell." Broken glass was supposedly found ten miles away, and the lighthouse keeper on the Isles of Shoals located seven miles off the coast of Portsmouth, New Hampshire heard the blast and thought it was an earthquake.*

The *Lowell Sun* also reported:

> *The scenes at the hospitals beggar description. The mutilated forms of men, women and children were taken to the hospitals as fast as they could be picked up at the scene of the disaster. The agonized cries of the men, women and children were echoing throughout the hospital halls and corridors, and the bruised and bleeding forms of little children were especially hard to look upon. Mrs. William Riggs, who lived next door to the powder magazines, was killed instantly.*
>
> *Additional policemen and the Massachusetts State Militia were called in because of looting. The homes were destroyed and abandoned and open for anyone to pilfer.*
>
> *Roughly 75,000 people came to look at the immense hole left in the ground and to contribute to the disaster relief fund. The money raised from donations was used to provide food and homes to those adversely affected by the explosion.*
>
> *People became afraid of the U.S. Cartridge Company's powder magazines off Andrew's Street near the company's mill. But apparently, talk of moving that storage magazine died down.*
>
> *Initially people thought the explosion was caused by a spark from a shovel or from a horse's shoe hitting a rock and possibly creating a spark. But the truth came out just after the explosion when Dr. John P. Torrey found a workman who knew what happened. Nine workmen were inside the powder house when they saw smoke coming from the floor.*

> *One left and brought back water to douse what they thought was a fire, but water had no effect. The men then fled from the powder house and seconds later, the witness felt himself thrown to the ground. The next thing he remembered were people standing over him.*
>
> *The cause was nitroglycerine that had leaked from the dynamite sticks casing and had soaked the floorboards of the powder house. It was the same soaked floor that the U.S. Cartridge Co. carpenter had been scheduled to replace.*

It was estimated that by today's currency the damage amounted to roughly $20,000,000.

Topsfield

The town of Topsfield built a powder house in 1818 in which to store its military stores. There is no mention of when it ceased to exist. In 1841, Topsfield built an armory/powder house for the "Warren Blues," a new military company formed in 1836. No record exists as to what became of this second building.

Walpole

Walpole's first powder house was built in 1773 on "Widow Robbin's high hill—now called Powder House Hill," so explains the town history. "It measured six feet square and cost £4.10. It was sold in 1787 by the town and later destroyed."

A second powder house was approved at the town meeting held on March 4, 1811. It was constructed of brick and was built later in the same year. The town history further reveals that the powder house "stood on a little hill off Elm Street about where the Walpole Freight Office later stood in 1925." This powder house measured 7 feet one way, and 8 the other. It survived until about 1840.

WARWICK

At an adjourned town meeting in May 1817, the small town of Warwick voted to build a powder magazine to secure the town's supply of powder and ammunition. It was voted to build the powder magazine "8 feet square, with well-burnt bricks lain in lime mortar, the walls to be 7 feet high, with a square roof well boarded and shingled, and to be ceiled [sealed] with good white pine boards, planed on the inside, and a good double floor well nailed; the door to be double, with a good lock and key; the walls to be 8 inches thick; and the foundation of flat stones to be built in a workmanlike manner." William Cobb, Caleb Mayo and Amos K. Whitney were chosen to supervise the construction of this powder magazine. Records do not indicate where the powder house stood or what became of it.

WATERTOWN

It is known that Watertown had a powder house in existence as early as 1777 because Thomas Staleford was assigned to guard it for six months during the Revolutionary War. Also, John Tainter of Watertown was put on guard duty at the powder house between October 1777 and April 1778 and from July 1778 to July 1779. No other information was available.

According to the *History of the Watertown Arsenal*, by Alan R. Earls:

> *In 1816 Watertown became the site of a major American arsenal originally comprising 40 acres of government-owned land situated upon the northern shore of the Charles River. The United States Army used this facility for the receipt, storage, and issuance of ordnance. In this role, it replaced the earlier Charlestown, MA arsenal. The arsenal's earliest plan incorporated 12 buildings aligned along a north-south axis overlooking the river. Buildings included a military store and arsenal used to store gunpowder, ammunition, and cannons, as well as other brick buildings with slate roofs designed in the Federal style used for shops and housing for officers and men. By 1819 all buildings were completed and occupied and a high wall enclosed the original quadrangle site.*
>
> *The arsenal's site, duties, and buildings grew gradually until the American Civil War, enlarging beyond the original quadrangle. During the Civil War it was greatly expanded to produce field and coastal gun*

Illustration of the large powder house/magazine at the Watertown arsenal, as drawn in 1853 for *Gleason's Companion Magazine. Matthew E. Thomas collection.*

carriages, and the war's impetus led to the quick construction of a large machine shop and blacksmith shop and a number of smaller buildings.

Activities and new construction continued to expand the size of the arsenal into the early 1890s. The activities changed significantly in 1892 when Congress authorized modernization to gun carriage manufacturing. During this period the arsenal became a manufacturing complex rather than a storage depot. A number of major buildings were constructed that reflected typical industrial facilities rather than the earlier buildings built in the Federal style of architecture. In 1897 an additional 44 acres were purchased by the U.S. Government and added to the site, and a hospital built to accommodate the large work force.

During World War I the arsenal nearly tripled in size. Building #311 was then reported to be one of the largest steel-frame structures in the United States, used to accommodate both large gun carriages and the equipment used to construct them. Railroad tracks ran throughout the arsenal complex. World War II brought an additional 7 acres with existing industrial buildings, as the arsenal was used to produce steel artillery pieces.

In 1968, the U.S. Army ceased operations at the arsenal; forty-five acres were sold to the Watertown Redevelopment Authority, while the remaining forty-eight acres were converted to the United States Army Materials Technology Laboratory in 1985. In 1995, all army activity ceased, and the remainder of the site was converted to civilian use. The Watertown Arsenal was added to the National Register of Historic Places in 1999.

This powder house still exists as part of the present-day Arsenal Mall.

Wayland

All that is known about the Wayland powder house is that it was built in 1772 and stood on the west side of the Sudbury River.

West Bridgewater

Very little is known about the West Bridgewater powder house that stood on the southwest corner of Powder House Cemetery on Matfield Street. Apparently it was in existence before the town built a powder magazine circa 1812–23 in the northwest corner of the upper gallery of the 1731 Double-Decker Town House that was torn down in 1823. Interestingly, the town's supply of ammunition was brought from the powder house to this magazine in the upper gallery, much to the delight of the boys, who found ample supplies of "plummets," which graced their writing books during winter school.

Upon the division of Bridgewater into different towns, the remaining military stores were distributed among the four towns. Whatever happened to the old powder house is unknown.

Westborough

The powder house at Westborough was built in 1818 and lasted until 1849. It stood in the cemetery, opposite the town hall. No further information is available.

Westminster

At a Westminster town meeting in March 1817, the town voted to build a powder house "for the purpose of keeping the Town Stock of Ammunition, etc." A committee was chosen to construct the powder house, but this action was subsequently annulled. However, in 1820, the matter came before the town again, when a similar vote was passed, and Alexander Dustin, Esquire; Deacon James White; and Luke Bigelow were made a committee to consider and report on the size, material and cost of such a house.

They reported in favor of one "six feet square, to be built of brick, with walls one foot thick, five feet eight inches high, a square roof of plank well shingled, with suitable shelves in the building, a suitable door with lock and key well hung, all thoroughly finished and the wood work painted at a cost of not less than 25 dollars not more than 30 dollars." The report was accepted, and Captain Nathan Whitney, Deacon James White and Edward Kendall were chosen as a committee to have charge of the work. They accepted the auction bid of Farwell Cowee, who erected the structure for $26.50, with the committee locating it at their own discretion on a large rock near the dwelling house (in 1893) of George Hager. The powder house stood for many years as the symbol of a civic and military landmark once very typical around New England. When it disappeared remains unknown.

Weston

Weston built its small square brick powder house in 1802 on the northeast corner of the Central Burying Ground near the intersection of Linwood Avenue and the Post Road. It was later purchased in 1834 by Mr. Alpheus Cutter and demolished by him for the bricks.

Westport

A wooden powder house was built on the Town Green near the east town landing at the village of Head of Westport circa 1813. It was used as an ammunition house by the Home Guard. In 1972, the Westport Historical Society moved the structure—six by six feet square and seven and a half

Left: This wood-shingled 1812 Westport, Massachusetts powder house was relocated in 1972 to the head of Westport Town Common. *Author's photo.*

Below: Moving the Westport powder house in November 1965. It was restored and dedicated in 1972 in a section of Westport called Head of Westport. *Photo courtesy of Westport Historical Society.*

feet high, with a wood-shingled hip-roof—to the center of the town green triangle to preserve it.

Woburn

Woburn town records state that the town voted to build a powder house on December 23, 1773 and chose a committee of three—the selectmen—to see that the work was done. What became of this first powder house is unknown.

In 1812, the town voted to build a second powder house on Powder House Hill. No further information is known.

Worcester

All that is known about the Worcester Powder House is that it was mentioned in September 1774 pre–Revolutionary War records and newspaper accounts as a target for British troops acting on orders from General Thomas Gage, royal governor of Massachusetts, to capture the gunpowder and ammunition "at the storehouse in Worcester." Gage was surprised by the strong reaction of the Massachusetts populace after the successful, unexpected British capture of gunpowder and ammunitions at the Charlestown (now Somerville) Powder House on September 1, 1774. As a result of this angry Patriot response, Gage decided to cancel another planned British army expedition to Worcester to capture the gunpowder in the storehouse there. According to published records, Worcester supposedly held the largest supply of gunpowder and ammunition in Massachusetts in 1774—second only to the supplies kept in the Charlestown Powder House.

Massachusetts had several fortifications protecting its extensive Atlantic Ocean coastline dating back into the 1600s. Many of these old forts presumably contained powder houses that are unfortunately not specifically documented.

CHAPTER 5
NEW HAMPSHIRE POWDER HOUSES

New Hampshire—the Granite State, whose straightforward state motto is "Live Free Or Die"—joined the Union as the ninth and deciding state to ratify the newly proposed Federal Constitution on June 21, 1788, thus officially making the Union a new Republic. Since its founding in 1623, New Hampshire citizens have been well known for their fiercely independent nature and distrust of big government. New Hampshire was the first colony to declare its independence from Great Britain, on January 5, 1776—a full six months before the Continental Congress meeting at Philadelphia, Pennsylvania, adopted the Declaration of Independence on July 4, 1776.

Another example of this fierce sense of independence in New Hampshire manifested itself as early as 1734, when a group of feisty residents residing in the present-day towns of Exeter, Brentwood and Fremont—the latter originally called Poplin—rioted against the Mast Tree Law, which forbade landowners from cutting down certain-sized white pine trees. These special trees were identified by a broad arrow mark cut into the tree by appointed British agents scouring the New Hampshire woods. These marked trees were to be cut and reserved for sole use as ship masts to be shipped to and constructed in Great Britain, for use in the King's Royal Navy. Many early New Hampshirites resented this British law because it prevented them from using the best timber on their own private property and from deriving the potential income from the sale of these valuable trees. In April 1734, local colonists gathered together and threatened to kill Lieutenant Governor David Dunbar, also the surveyor general of the King's Woods in New Hampshire,

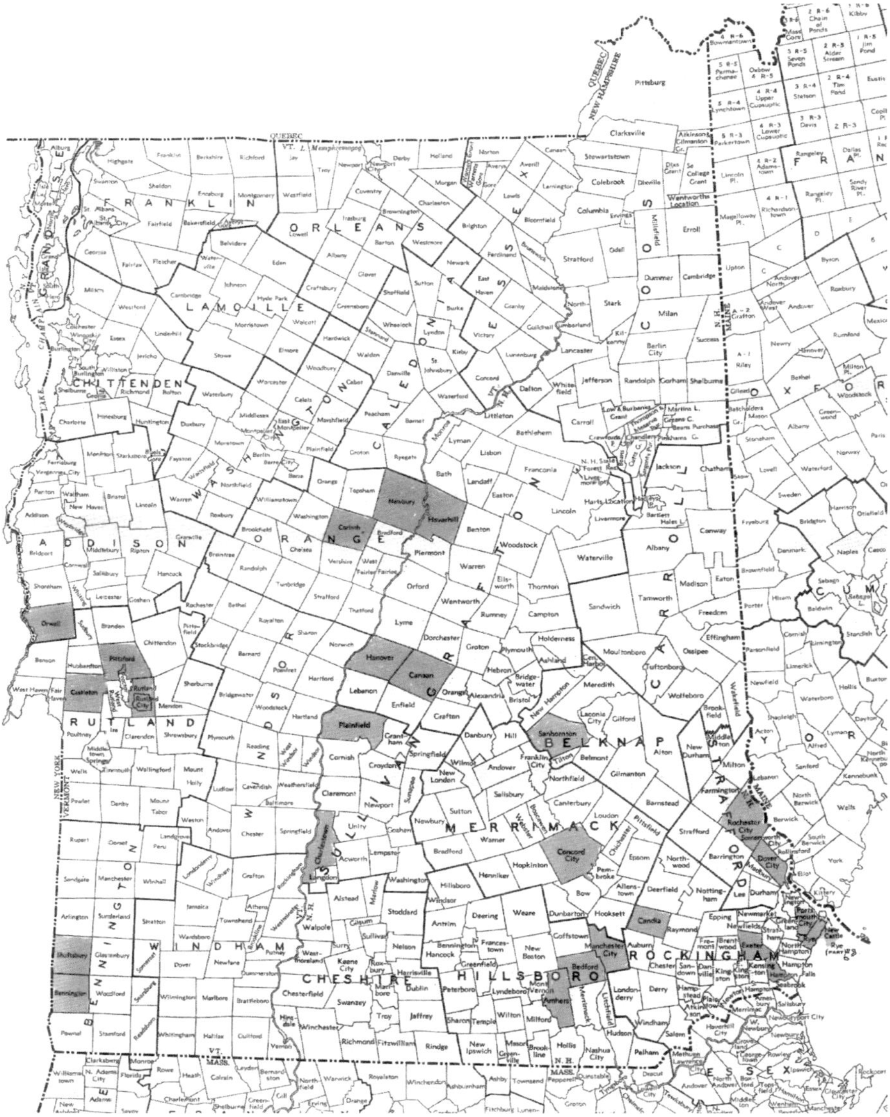

Map of New Hampshire and Vermont highlighting cities and towns with powder houses.

when he arrived to inspect illegally cut white pine trees at Copyhold Saw Mill. This mill was located close to the present-day town line of Brentwood and Fremont, both of which were, at the time, part of Exeter, New Hampshire. Dunbar fled back to Portsmouth in fear of his life. Like future rebellions against the British-imposed Stamp Act and Sugar Act of the 1760s and the Tea Act and Quartering Act of the 1770s, this became the first revolt of its type against the unfair British-imposed Mast Tree Act of 1708. The Mast Tree Riot of 1734—along with the 1772 Mast Tree Riot at Weare, New Hampshire—combined with the various other oppressive British acts to eventually lead to the cry by American colonists at the outbreak of the Revolutionary War of "no taxation without representation in Parliament!" Therefore, it is not unreasonable to say that because of the Mast Tree Riot of 1734, the first true shots fired for American Independence echoed from the wooded countryside surrounding the Copyhold Saw Mill.

In December 1774, the first overt act of the American Revolution took place at Fort William and Mary on the tiny seacoast island town of New Castle, located two miles outside of Portsmouth, New Hampshire. On December 13, 1774, Paul Revere made the first of two famous "midnight" rides, the first being to Portsmouth to warn residents that the British army was coming from Boston to take possession of the cannon, small arms and one hundred barrels of gunpowder stored at Fort William and Mary in New Castle. By capturing the military ammunition and supplies at the fort, the British hoped to prevent these military stores from falling into the hands of the increasingly hostile American Patriots. Revere's message enraged the locals, who then assembled four hundred strong and, on December 14, marched to the fort where they were fired on by the six British soldiers standing guard. No injuries were recorded, and the American Patriots then stormed the fort and captured the British soldiers and the military stores from the powder house. A portion of this captured gunpowder was presumably used at the Battle of Bunker (Breed's) Hill on June 17, 1775.

During the War of 1812, the only naval battle to take place in New Hampshire occurred off the coast at a place called Ragged Neck in Rye Harbor. A British barge was seen pursuing a fishing schooner into Rye Harbor in May 1814 when the British fired at the schooner. Because the Americans had been fearful in recent days of an impending coastal British attack at Rye or Portsmouth, the local militia was waiting behind stone walls and fired at the British barge. The battle, though short, was intense while it lasted, and supposedly, the coxswain on the British barge was seen to fall and was believed to have been killed. By the end of the battle, most in the

Rye militia had used their entire supply of ammunition to chase the British from Rye Harbor.

For the record, Stratham, New Hampshire, is the only known town in the state to record within its town records the decision not to build a powder house, with that vote taking place in 1809.

Amherst

The powder house in this very historic old New England town stood on the easterly side of the burying ground behind the 1823 County Courthouse on Main Street. The construction of the powder house was approved by a town

Illustration of the 1774–75 wood-plastered Amherst, New Hampshire powder house that stood in the village burying yard until about 1845, when it was removed. Illustration from *Colonial Amherst*, by Emma P. Boyleston Locke (1916).

vote on September 15, 1774, and it was eventually built of chestnut logs, hewed twelve inches thick and lathed and plastered on the outside. It was ten feet square with a pointed roof. Paul Dudley and Timothy Smith were appointed to build the powder house. It performed its duty for about seventy years and was taken down about 1845.

BEDFORD

Built circa 1924, this powder house, though not typical of the usual powder houses described within this book, was part of the DuPont Gunpowder/ Dynamite Manufacturing Company and was located one-half mile south from the Daniel Webster Highway, almost directly in back of the formerly popular Bedford Grove roller skating rink. The author felt it important to include this incident within this book because of the extensive devastation and fear that this explosion wrought on the Bedford and Manchester communities.

As reported in the *Manchester (NH) Union Leader* on July 4, 1944:

> *About 15,000 pounds of dynamite (40-60 mixture) was stored in one of two small, tin-covered wooden buildings that served as powder houses for the DuPont Company. Shortly before midnight on the night of July 3rd a terrific explosion rocked the surrounding area for miles leaving a huge crater about 20 feet in depth and 70 feet in diameter. This massive explosion of 7½ tons of dynamite was felt as far south as Lexington, Massachusetts and as far west as Worcester, Massachusetts and created massive pandemonium throughout the Bedford, Manchester, and Merrimack communities. Thousands of men, women and children, many in their pajamas and night coats, scurried onto Elm Street in Manchester and adjacent streets some crying, screaming, and others hysterical, not quite sure what had happened, what to do, or where to go. Children huddled in fear about their parents. Thousands stumbled over vast amounts of shattered glass from broken storefront and apartment windows, as well as fallen bricks, and other debris that littered sidewalks and streets caused by the terrific concussion throughout downtown Manchester and neighboring communities. In some instances, produce and other groceries on display in local markets rolled out onto downtown streets. It was estimated that $25,000 worth of broken glass was done in Manchester alone. Though no one was killed, many were injured by flying glass, others from falling bookcases and wall decorations,*

> *and others were thrown down to the floor or against walls and furniture during the blast. In one case, a girl who was roller-skating at the Bedford Grove Roll-a-Way was grazed on the shoulder by the huge crystal ball that threw reflected light over the rink after it fell from the ceiling onto the floor. Overhead lights and Chinese lantern decorations crashed all over the rink as several skaters jumped out of windows with their roller skates on, while outside spectators dashed to safety across the highway and then watched the building shake and the roof split. The explosion was so great that it sent a great pressure of air against windows and buildings throughout the area and shook the ground to the extent that buildings shook fiercely and many household items were knocked from their shelves.*

Since this explosion happened during World War II, many area residents were unsure at first if it had been a bomb attack perpetuated by the German or Japanese enemy. Throngs of people tried to escape the city for fear of additional bombings, while others drove to the vicinity of the explosion to survey firsthand the seemingly unimaginable damage. Traffic was so thick to Bedford that a police car with siren blaring had to snake its way along the main thoroughfare. Local police were virtually helpless at first in calming the overwhelmed citizenry and had to call in the New Hampshire National Guard and the Grenier Field Military Police to help restore order and to calm fears. Witnesses to the explosion in Bedford stated that the tremendous blast created a geyser of flame hundreds of feet in the air, and there followed a series of smaller blazes high in the clouds in the manner of fireworks. No one was naïve enough to think that the detonation was any kind of fireworks, however, as expressions made on hastily filled streets indicated that it might have been a bomb dropped from an army plane or the gas storage plant on lower Elm Street. The events of the night were a wild climax to what people thought was to be one of the quietest nights before the Fourth of July that Manchester had ever witnessed. Instead of a feeling of jubilation, there was only an air of terror. The sudden, shattering blast, the origin of which was not known for several minutes after it occurred, created a mysterious feeling of awe, which was worse than knowledge of the exact cause would have created.

When the facts became known, automobiles by the hundreds started for the vicinity of the Bedford Grove, some anxious over the fate of relatives who were known to be attending the midnight dance at that resort. Others were just curious, creating a traffic snarl that resulted in Bedford Police calling for reserves from nearby police departments.

Upon investigating the powder house explosion, it was determined by the Bedford police chief that the explosion was no accident. Kerosene-soaked overalls covered with mud and muck from the woodlands were found along a path some distance from the crater that resulted from the explosion.

It was reported by still another source that investigation showed someone had been tampering with the second powder house, which contained detonation caps, apparently with the use of a crowbar. It was also revealed that a thorough search was made of the second powder house for fingerprints.

Needless to say, this exceptional powder house explosion on the eve of July 4, 1944, during the midst of World War II created such fear, hysteria and injuries that it should never be forgotten in the annals of New England history. It was a miracle that no lives were lost and unfortunate that the culprit(s) were never captured.

CANAAN

An artillery company was organized in Canaan in 1820. The militia system of New Hampshire at that time was doing its best to make citizen soldiers of nearly every man. The local artillery company had all arms but one—it lacked a brass six-pounder gun. A group of interested local citizens gathered together in the hall of Captain Joshua Harris's Tavern on Canaan Street and voted to form an artillery company and appointed a commission to make an application to the state, asking to loan them a cannon. The application to Governor Samuel Bell was successful, and orders were sent to purchase land and build a "gun house" on it. Captain Harris gave the state a deed of land, and the gun house was built on it. For the next eighteen years or so, the artillery company patriotically hauled the cannon up and down Canaan Street on special occasions, firing off blank cartridges filled with green grass. The Canaan Artillery ceased operating circa 1840, and the state conveyed its title to the gun house property to the town.

CANDIA

The Candia Gun House formerly stood on High Street prior to 1845, when the militia's cannon was removed from the gun house and brought to nearby Raymond, New Hampshire. In 1858, the cannon, which had been the pride of the artillery of the Seventeenth Regiment for forty years, was sold for junk.

Charlestown

There is scant information that a powder house once stood within the walls of Old Fort #4 that originally protected the western frontier of New Hampshire. It was constructed about 1744–46 and saw significant military activity during the French and Indian War between 1755 and 1763. The fort fell into disrepair by the late 1770s and gradually disappeared. The fort was reconstructed during the mid-twentieth century and is now a popular New Hampshire living history museum.

Charlestown

In 1807, the Town of Charlestown gave the State of New Hampshire thirty square feet of land located in the rear of the South Meeting House for the purpose of erecting a gun house. It was still in use as late as 1860, when it was used by the Sixteenth Regiment.

Concord

Concord serves as the capital of New Hampshire and was granted in 1725 as Rumford. The name was changed to Concord in 1765 when the town became incorporated. Concord became a city in 1849.

All that is known of the Concord Powder House is that it was constructed on stilts "on the Powder House Lot." Town records do not relate the year of construction or when the structure ceased existence. This is the only known powder house in New England that was built on stilts.

DOVER

Dover was the site of the first permanent settlement in New Hampshire, founded in 1623. It originally consisted of the current towns of Durham, Lee, Madbury, Somersworth, Rollinsford and a portion of Newington.

According to *Port of Dover: Two Centuries of Shipping on the Cocheco*, by Robert A. Whitehouse and Cathleen C. Beaudoin, "A Dover Powder House [was] in existence as late as 1877 standing along the banks of the Cocheco River where workmen were about to dredge a channel at a point near the Powder House." No further information could be found.

DOVER

Dover voted on March 14, 1809, to "purchase a lot of land off Captain Moses Ham to set the Gun House on if he would sell for $30.00 which he agreed to do at Town Meeting."

The town (now a city since 1855) had tried to build a gun house in 1808 on a lot near the courthouse, which caused much opposition from adjoining property holders, so it was not built at that site.

EXETER

Exeter was founded in 1638 by the Reverend John Wheelwright, who left Massachusetts to escape religious persecution. It is the third-oldest town in New Hampshire and originally embraced the current surrounding communities of Newmarket, Epping, Brentwood, Poplin (now Fremont) and South Newmarket (now Newfields).

Exeter served as the Revolutionary War capital of New Hampshire between 1775 and 1782. Exeter's claims to fame include approving the first written constitution of any of the original thirteen colonies, which declared the colony of New Hampshire independent of Great Britain on January 5, 1776—a daring yet patriotic act performed a full six months before the Continental Congress at Philadelphia issued its own Declaration of Independence on July 4, 1776. Exeter is also home to the prestigious Phillips Exeter Academy, founded by John Phillips, incorporated in 1781 and opened in 1783.

The 1771 Exeter powder house contained gunpowder captured by the colonists from the British-held Fort William & Mary in New Castle, New Hampshire, in December 1774. It is believed that some of this gunpowder was later used at the June 17, 1775 Battle of Bunker (Breed's) Hill that could actually be heard in Exeter as the battle raged forty-five miles away. *Author's photo.*

The Exeter powder house was built in 1771 on a small point of land variously known through the past 350 years as Powells' Point, Rundlets' Point, Duck Point and, since 1999, Powder House Point located on the south end of Jady Hill Avenue. It can easily be seen across the Squamscott River from Swasey Parkway or by hiking out to it along a path located at the south end of Jady Hill Avenue.

This brick powder house was originally intended to be built as a round structure, but artisans of that time decided it would be easier to build in the square. They also wisely placed the door on the south side of the building, therefore exposing it to more constant sunlight and thus preserving the wood longer and making it more accessible from the river. The structure measures ten and a half feet wide per side, with each side facing a cardinal direction. It is eight and a half feet high from ground to eaves, with the whole structure measuring eleven feet seven inches high from the floor to the apex of the inside roof measurement. Brick walls, 16 inches thick, rest on a base of granite extending about six inches above the ground. The sixth course of brick from the base is of headers, projecting out from the wall about one inch, giving a finished appearance, and above this, every fourth course is of headers. It has a pyramid-shaped roof, bricked and boarded over and shingled with the upper three rows of brick corbelled outward three-fourths of an inch on each row to meet the eaves. Ventilation shafts are located nearly in the middle of all four walls, purposely designed at an upward right angle so that wind and rain cannot pass through the passage. It had a strong wooden door, covered with sheet iron, 4'9" by 3'3/4" thick and ornamented with hand-wrought hinges. Though it is not the original door, it is still an old door at approximately 135 years old. The double wood floor is about one foot above the ground and the space underneath filled with fieldstones so that the inside is always dry. The bricks, apparently of local make, vary in length from 7 to 7¾ inches and in depth from 1½ to 2 inches.

As a young man growing up in Exeter during the early 1970s, this author sadly remembers hiking out to this ancient landmark one autumn day and viewing the severe vandalism inflicted on it. Despite having been repaired by the Exeter Conservation Commission in 1969, by the early 1970s, bricks were again missing from above the powder house door, the door was partially off one of its hinges, the door lock was missing, beer bottles and trash littered the interior, shingles were torn from the roof and graffiti had been spray-painted on the outside walls. Fortunately, determined residents of Exeter again restored the powder house in 1999 and 2000 by putting on a new slate roof complete with copper trimmings. Grey granite posts with black chain fencing were erected, and granite block steps were placed on the small knoll leading up to the powder house. A flagpole was erected with sensor lighting, and an information kiosk sign, granite benches and various flowering plants were all nicely placed to enhance the powder house site. A time capsule was also placed within the refurbished structure under the new wooden floor. All this was done without taxpayer funds. The preservation of this historic landmark is an exemplary example of the tremendous pride Exeter residents take in preserving its illustrious heritage. In recent years, there has been a resurgence of interest in restoring and preserving area powder houses, and as in Exeter, fine restoration work has also been done in nearby Portsmouth, New Hampshire, and Newburyport, Massachusetts.

The Exeter powder house played a small yet important role at the outbreak of the American Revolution. On December 14 and 15, 1774, an armed mob of four hundred local Patriots stormed Fort William and Mary in New Castle, New Hampshire, and captured six British soldiers guarding it despite the British firing off cannons and small arms to scare off the insurgents. The American colonists secured nearly one hundred barrels of gunpowder from the British-held fort as well several cannon and small arms. Some of the captured gunpowder from Fort William & Mary was later divided up and sent mostly to inland New Hampshire towns like Nottingham, Epping, Poplin (now Fremont), Brentwood, Londonderry, Durham, Kingstown (now Kingston) and Exeter for safekeeping from the more Tory-populated seacoast towns like Portsmouth, where just four barrels were hidden. Four of these one hundred barrels of captured gunpowder were later stored at Esquire Zaccheus Clough's remote Hook-In-The-River Tavern, which still stands at 125 South Road in historic Fremont. Another twenty-nine barrels were stored in various homes around Exeter, with some of it later being gathered up and placed within the Exeter powder house for safekeeping. A small quantity of this captured gunpowder was also stored by "Powder Major" John DeMerritt Jr. (1728–1826) in a secretly dug hidden chamber in the cellar of his home in Madbury. DeMerritt is said to have transported by ox cart his cache of powder just in time to be used at the June 17, 1775 Battle of Bunker (Breed's) Hill in Charlestown, Massachusetts. John DeMerritt Jr. was soon thereafter given his name "Powder Major" because of the significant role he played during the capture of Fort William and Mary and for carrying powder down to Bunker Hill. A boulder monument erected in 1929 marks the site of the Powder Major's home on Madbury Road in Madbury.

The 1774 capture of British-held Fort William and Mary made this New Hampshire event the first overt act of the American Revolution, with defiant American Patriots rebelling against British authority. It also set the stage for

Esquire Zaccheus Clough's circa 1753–57 Hook-In-The-River Tavern at Poplin, New Hampshire—now called Fremont—where barrels of gunpowder captured at Fort William and Mary in New Castle were hidden for safekeeping from British authorities. This old home pictured in 1885 was typical of several in adjoining southeastern New Hampshire towns that divvied up the one hundred barrels. *Matthew E. Thomas collection.*

the April 19, 1775 Battle of Lexington and Concord. After the successful capture of Fort William and Mary, there was great excitement throughout Portsmouth for three or four days, with large numbers of local Patriots streaming in from the outlying country towns ready to fight the British if it became necessary.

Samuel Hobart of Hollis, New Hampshire, established the first powder mill in New Hampshire at King's Falls off Powder Mill Road in Exeter in 1776. The gunpowder manufactured extensively here was used to supply ammunition to the various New Hampshire regiments serving during the Revolutionary War. After the Revolution, Hobart converted the powder mill into a nail manufactory. In the 1830s, it was again converted into a powder mill with severe explosions occurring in August 1840 and May 17, 1843, when a workman was literally blown to pieces and fragments of his body caught and hung in branches of surrounding trees.

HAMPTON FALLS

A document issued from the Town of Hampton from Captain Samuel James, dated September 17, 1813, requests "to warn and notify all the Non-commissioned officers & soldiers belonging to the Artillery company in your district to meet at the gun house in Hampton Falls on Monday the 27th day of September inst." This is all that is known about the Hampton Falls gun house.

HANOVER

The Hanover gun house was built circa 1836, but it was dismantled and removed to another section of Hanover by 1843. No further information is available as to whether it is still standing.

According to a petition letter addressed to the New Hampshire State Senate and House of Representatives and written by petitioners Joseph M. and Lucinda C. Givven dated June 7, 1843, the petitioners asked the state to permit them to purchase the lot belonging to the State upon which formerly stood the gun house. The gun house lot adjoined land belonging to Lucinda, who purchased her lot in 1837. Since that time, the gun house had been removed and relocated to another part of Hanover.

HAVERHILL

The Haverhill Powder House was built on Powder House Hill by the State of New Hampshire in 1812 of massive granite slabs about twelve feet square and was used during the War of 1812. Haverhill was then, as during the Revolutionary War, located at a geographical point of exposure along the Connecticut River, and troops were stationed for protection at this northern frontier section of New Hampshire. This powder house was a town landmark for three-quarters of a century before it was taken down and the stones used in the construction of a receiving vault at the cemetery on Ladd Street. In 1847, the State of New Hampshire deeded its interest in the powder house and the one-eighth acre it sat on to the Town of Haverhill. The 1888 Haverhill town history contains an 1863 photograph of Haverhill Village showing in the far distance the powder house atop Powder House Hill.

The Haverhill town historian lamented in 1888, "It was a shame that the town did not allow this ancient landmark to stand and be restored to its primitive state. I fear the town is not imbued with a sentiment for the past, as ought to inspire her, in view of her historic character. We have truly been iconoclasts."

In 1806, before there was a powder house on Powder House Hill, Josiah Burnham was publicly hanged for murder from this eminence before an estimated crowd of ten thousand spectators.

Manchester

The Manchester Powder House, built circa 1888–93, was another New England landmark that was owned by commercial interests rather than by the community. According to the August 20, 1903 edition of the *Manchester (NH) Mirror*, "It was located in a section of the city called *Dodgeville* in the vicinity of the old former Mammoth Mills Department Store and a comparatively short distance away from the Elliot Hospital. It was a small brick structure with an iron roof and heavy iron door built upon a small eminence on the east side of Mammoth Road on land formerly owned by a Robert I. Stevens at the time." It stood on Mammoth Road between Oakland Avenue and Candia Road. It no longer stands.

In 1903, the powder house was owned by the Oriental Powder Company of Boston, which used it to store dynamite, gunpowder and fireworks. The company at that time had an agent living in Manchester who supplied powder and dynamite to the Manchester Hardware Company and to other businesses needing such items within the city.

Residents living in Dodgeville and its vicinity were legitimately concerned in August 1903 for the safety of the neighborhood after a devastating powder house explosion took place earlier that summer at a similar powder house, killing and injuring dozens of people in Tewksbury and Lowell, Massachusetts. Though residents had been led to believe that the Oriental Powder Company only kept small quantities of explosives within the powder house, it was discovered during the summer by a neighbor living nearby that it was nearly full of powder, dynamite and fireworks. This caused considerable alarm to many Dodgeville residents, who planned to petition the mayor and aldermen of the city to abate the alleged nuisance. They were afraid a potential explosion of the powder house would flatten the entire

The 1808 powder house inside the former remains of Fort William and Mary in New Castle. Fort Constitution was built in 1808 and incorporated the original old fort. The first overt act of aggression against the British just prior to the Revolutionary War took place here on the night of December 14, 1774, when four hundred local patriots captured the fort and took cannons and one hundred barrels of gunpowder. Had casualties occurred during this 1774 raid, the "shot heard round the world" might well have occurred in New Castle, New Hampshire, instead of at Concord and Lexington, Massachusetts, five months later in April 1775. *Author's photo.*

neighborhood and possibly do harm to a nearby schoolhouse as well as to travelers using the electric railway that passed in front of the powder house. Another concern was that the powder house, with its iron roof standing on a small eminence, might be susceptible to a lightning strike, thus endangering the neighborhood in that manner or by potential mischief-makers playing around the building.

No further information could be found to determine what exactly happened to this powder house or what steps the city took to address the concerns of Dodgeville residents.

NEW CASTLE

Within the walls of Fort Constitution is a rectangular brick powder house, constructed as part of the fort in 1808. Fort William and Mary underwent reconstruction in 1794, 1798 and 1808. It apparently was in 1808 that Fort William and Mary was officially renamed Fort Constitution, though Secretary of War Henry Dearborn referred to it as such for the first time as early as 1801. After the American Revolution, it had briefly been called Fort Hancock in honor of John Hancock, the first signer of the Declaration of Independence. Physically, very little remains of Fort William and Mary except some old foundations and part of the south wall. It sits on the rocky northeast point of Great Island in New Castle, which was incorporated as the fifth-oldest town in New Hampshire on May 30, 1693. The existing powder house was used from 1808 until the end of the Civil War. In 1842, two additional powder houses were built within the fort, which have long since disappeared.

According to a report filed in 1740, an earlier powder house stood on this same site. At that time, it contained a half barrel of gunpowder. From this circa 1740 powder house, one hundred barrels of gunpowder were captured by four hundred angry New Hampshire Patriots who stormed the British-held fort on December 14 and 15, 1774. This raid upon Fort William and Mary is considered by many historians as the first overt act of the American Revolution.

Incidentally, British troops did arrive at Fort William and Mary on December 17, just three days after Patriots captured the fort. They arrived on board the HMS *Canceaux* with further reinforcements on the HMS *Scarborough* that remained to protect Governor John Wentworth, garrison the fort and restrict shipping to Portsmouth until the autumn of 1775. Governor Wentworth was later forced in August 1775 to flee onto the HMS *Scarborough* only two months after Patriots had placed a cannon in front of his mansion, forcing the governor to surrender his trusted friend John Fenton, who had defended the beleaguered governor in an unpopular speech, after which Fenton was chased through the streets of Portsmouth by an angry mob. The governor never felt safe in his home again. He fled with his family to Fort William and Mary and then onto the *Scarborough*, where he sailed to Boston, never to return to New Hampshire again.

According to information from various newspaper accounts printed in the July 5, 1809 *Portsmouth (NH) Gazette*; the July 8, 1809, July 29 and September 2, 1809 *Portsmouth Oracle*; the July 7, 1809 *Newburyport (MA) Herald*; and the July 8, 1809 *Boston (MA) Columbian Sentinel*, the following details of this tragic event were reported by a witness to the catastrophe:

Interior view of the powder house at Fort William and Mary. Note the small window in the rear of the powder house, a rare architectural feature. *Author's photo.*

Aerial view of Fort Constitution, formerly Fort William and Mary, on the island community of New Castle, New Hampshire. *Matthew E. Thomas collection.*

During the midst of a 4th of July celebration, between 4 and 5p.m. two chests of powder and a number of loose cartridges that had been laid out in preparation for firing a salute, took fire and in the explosion killed several soldiers and four young civilians. Several others were wounded in addition to causing much other essential damage. The quantity of powder exploded was between 3 & 400 lbs. in weight. The deaths caused by this melancholy event (which have already come to our knowledge) are right, and it is feared, from the burnt and mangled situation of several others, that they cannot long survive.

Among the killed and wounded are the following: Ephraim Pickering, Esquire of Newington, NH dead! A son of Stephen Paul, of Kittery, [ME] *aged 14, dead!—these two had just landed to see the new fort. John Mitchell, a lad, and James Trefethen, a lad, both of New Castle, dead! One young man who was a spectator about 15 years old, his name unknown, dead! Sergeant Albert* [or Abbot] *dead! Patrick (or Peletiah) McDaniels, dead! Theodore Witham, of New Gloucester,* [ME] *dead! Samuel Stevens, Reo Gamoth, Gideon Gold, Edmund Hurd, John Ricker, Robert Miller, the wife of Captain John Walbach and some others whose names we could not learn, all wounded. Several of whom it is feared, mortally.*

No blame whatsoever can be attached to Capt. [John] *Walbach commander of the Fort. He directed an under officer to attend carefully to the firings of the afternoon, while he should enjoy himself with a few select friends at his house, which is within the Fort, and also within thirty-feet of the awful explosion! Judge ye of the surprise of this excellent officer, when his house was nearly blown up, with his guests and family with him! The windows,* [along] *with part of the house bursting in and everything breaking and crushing around them, so that there was scarcely a pane of glass, or an article that could be broken, but was dashed to atoms! The Captain's lady was knocked down, but not mortally hurt, while the Captain and his guests escaped with but little injury. But the Fort and yard presented an awful and tragic scene legs, limbs, bones and human flesh were scattered in various directions, too shocking to attempt description, and such as my eyes never before witnessed, and I pray to Heaven never may be called to again.*

Yours respectfully,
C.P. (Reporter)

Another account later states, "Robert Miller, of Dunbarton, NH, a private in the U.S. [military] service, was, with other soldiers, killed by the explosion of the [powder] magazine at Fort Constitution in Portsmouth [actually in New Castle]." The account goes on to state that "on this occasion the severed leg of one of the unfortunate men was thrown a distance of several rods with such force as to break through a door of the house occupied by Colonel [Captain] Walbach, who commanded the garrison [Fort] and the United States troops in that vicinity."

Samuel Stevens succumbed to his injuries on July 21, leaving a wife and six children to lament his loss, and Gideon Gold (or Gould) also died later in the summer, bringing the total number of Independence Day explosion victims to eleven. This was most likely the worst Fourth of July explosion ever to take place in New Hampshire history.

There were originally two powder houses that stood within Fort Constitution. The larger of the two still stands and was built of brick in 1808. It stands ten feet high and eight feet wide and is rectangular and capped with a shingled hip roof. An archway door ornamented with brick headers is the only opening to the building. Two columns of tapering bricks provide added strength and support to the structure's east wall up to the eaves of the roof. The other wall is strengthened by being placed against a large fortified granite wall located on the inside of the fort. The actual powder house from which New Hampshire Patriots captured one hundred barrels of gunpowder in 1774 was unfortunately destroyed in the 1809 July 4th celebration explosion.

Plainfield

The Plainfield Gun House stood near the Plainfield Village Cemetery at an unknown date. It was used to shelter the cannon used by the local militia. Another gun house formerly stood just to the east of the Meridian Congregational Church.

Portsmouth

Portsmouth was founded in 1623 as "Strawbery Banke," a name it held until it was changed to Portsmouth in 1653. This town became a city in 1849. Portsmouth is the second-oldest town in New Hampshire and served as the Provincial capital of the colony until 1775, when the seat of government was moved to Exeter where fewer British sympathizers lived. Portsmouth is located at the mouth of the Piscataqua River and is a popular tourist destination along the New Hampshire seacoast, noted for its impressive array of outstanding architecture, including such varying styles as First Period, Georgian, Federal, Greek, Gothic, Romanesque and Queen Anne Victorian structures. Portsmouth is home to Strawbery Banke—a living history museum noted for its collection of restored and preserved early American homes, shops and taverns dating back to the seventeenth, eighteenth and nineteenth centuries.

The earliest Portsmouth powder house originally stood in or very near the North Cemetery off Maplewood Avenue. It is believed that this original structure dated back to the 1750s or '60s. It was demolished soon after a new powder house was built on Islington Street in September 1811, just in time for the War of 1812 to break out. This new powder house stood closer to the "Plains," which had been the site of annual militia trainings for many years between 1720 and 1851.

The Portsmouth powder house is the largest structure of its kind in New Hampshire. It is round and made of bricks, with a shingled conical roof. It stands roughly seventeen feet high and twelve feet wide and is ornamented with raised brick dentils circling under the eaves of the roof. The walls are five bricks thick, interspersed with air vents. The tall wooden door is studded with long square nails to prevent it from being broken into with axes. Two long, strong, hand-wrought iron hinges support the door. The door frame is capped by brick headers. Unlike other area powder houses, this one has a smaller wooden door with two hand-wrought iron hinges located above the ground-level door, which allowed kegs of powder to be hoisted through for storage on the second floor. This small wooden door is also capped with a slightly arched row of brick headers. Between the two doors is an ornate marble plaque that gives the date of construction as September 1811. There are several ventilation shafts, located mostly on the south side of the structure, that prevent wind and rain from entering the interior. This powder house was built by Captain Daniel Marden, a master mason who built it on a stone foundation on a rocky ledge. Tradition has it that bricks, wooden trunnel pegs and beams from the old North Cemetery powder house were incorporated into the new 1811 structure. Like so many other remote or hidden powder houses, this structure has also been subjected to vandalism and spray-painted graffiti over the years. This powder house was used during both the War of 1812 and Civil War. Until 1904, it housed explosives for the city, until another powder magazine, which has long since disappeared, was built on Jones Avenue.

For many decades, this powder house was obscured behind a massive city water tank erected in 1892 and dismantled in 2006. The powder house and lot were restored in 2007, and the site was renovated with a brick walkway, sitting benches, lighted lamps and flowering bushes.

Until the restoration of the Portsmouth powder house, it seemed to have been an ignored and forgotten piece of Portsmouth's colorful and historic past. Few Portsmouth residents knew of its existence, and local history books gave it scant, if any, mention at all. It now sits on a small rocky knoll, proudly displaying its newfound visibility and providing ample testimony to the historic role it has played in Portsmouth's illustrious past. It can now finally take its rightful place alongside other notable Portsmouth landmarks. The citizens of Portsmouth are to be congratulated for recognizing the importance of and preserving this rare New Hampshire landmark.

An interesting story about the 1811 Portsmouth powder house appeared in the June 3, 1876 *Nashua (NH) Weekly Telegraph*, which stated, "The Portsmouth Powder House has been undergoing repairs. Inside is stored a quantity of powder, belonging to no one knows whom, and cartridges prepared in the War of 1812 on which are written: 'This one is for the heart of General Hull,' 'this one for the heart of General Gates, etc.' These missiles seem not to have performed their work, but have slept in quiet solitude for over 60 years."

A one-story brick gun house was built in 1808 near the intersection of South Street and South School Street, near the South Mill Pond. It was used for many years as a New Hampshire State Arsenal to house three twenty-four-pounders and three eighteen-pounders on traveling carriages, in addition to military supplies. The second story was added sometime after the Civil War when the arsenal was no longer needed for military purposes. A wall that enclosed the gun house was removed before the 1890s. This structure was demolished in 1920 after a new state armory was built in 1916.

Opposite, top: One of the largest powder houses still standing in New England, this imposing 1811 circular brick structure in Portsmouth, New Hampshire, replaced an earlier powder house formerly located in the historic North Cemetery. Note the unusual second-story door to receive barrels of gunpowder. *Author's photo.*

Opposite, bottom: Circa 1931 photograph of the 1811 Portsmouth powder house hidden behind the old 1892 water tower. The water tower was removed in 2006 and the entire powder house lot restored and beautified into a small park. Photo from Lost Examples of Colonial Architecture, by John Mead Howell. *Courtesy of Portsmouth Athenaeum.*

Rochester

The Rochester gun house, built circa 1792, stood near a schoolhouse on the east side of the town common, bearing the inscription "Gun House, 39th Reg't. The Thirty-ninth Regiment had an exceptionally fine brass six-pounder, which was returned to Portsmouth when the regiment disbanded in 1849.

Rye

The coastal New Hampshire community of Rye built a fourteen-foot-long, ten-foot-wide gun house on town-owned land near the town pound in 1821. It was located just east of the Christian Church. In 1853, the old gun house was sold at auction to Orin Drake for four dollars.

Sanbornton

According to the Sanbornton town history, the building of the circa 1809 powder house was attended to with considerable legislative formality, as the following record attests:

> *November, 1808; first Friday. Town voted to build a powder house, and chose a committee of three—Bradstreet Moody, Jeremiah Tilton, and Nathan Taylor—to make draft of the same, estimate cost, and receive proposals, reporting at the next annual meeting. March 18, 1809. Powder house committee reported that it stand on the top of a ledge hill, on the easterly part of Esq. Lovejoy's land: (be) round: eight feet in diameter on the inside; eight feet high; and then topped off in the fashion of a cone. Material, stone or hard-burned brick, laid in, and iron hinges, and pointed with lime on the outside. Door* [made] *of white-oak plank, well hung, with iron hinges, and secured with a good lock. Job to be bid off during the auctioning off of the poor; but not to go over $60.00 there being offers to do it for that amount. Thickness of walls, time of contract, and lightning spire (whether this year or next) (are) to be left discretionary with the selectmen. Report accepted.*

Existing town records do not explain what became of the Sanbornton Powder House.

CHAPTER 6
RHODE ISLAND POWDER HOUSES

Rhode Island—the Ocean State—defiantly joined the Union as the thirteenth state on May 29, 1790. It was the last of the original thirteen colonies to adopt the U.S. Constitution and when it did, it demanded that a Bill of Rights be added to that venerable document. Though it may be the smallest state in the nation, it has one of the most colorful, defiant and determined histories of any of the fifty states. This tiny state was not about to be bullied by its larger neighbors or by the newly established Federal government under the 1781 Articles of Confederation that loosely tied the thirteen colonies together as a political body. Consequently, the state conducted its political affairs in a manner that looked out for the best interests of the people of Rhode Island—like the sovereign state that it was—not out of a need to be selfish, but out of necessity to survive and overcome its economic challenges.

Rhode Island had suffered immensely during the Revolutionary War from the devastation caused by the three-year British occupation of Newport, as well as several British raids and assaults on various towns and cities during the Revolution. It also suffered much from the high rate of inflation in the 1780s that followed the conclusion of the Revolutionary War. So painful was this high inflationary period in American history that it caused riots in other New England states. Much of Rhode Island's independent attitude came from its various objections against ratifying the proposed U.S. Constitution. One of its concerns was over who should have the authority to levy import taxes on Rhode Island citizens—the sovereign state of Rhode Island or the new proposed federal government? Rhode Island citizens felt strongly that only its own legislative body should be able to levy import taxes—after all, had not the state fought a revolution against British trade restrictions and import duties just to hand over that same power to a federated Congress over which Rhode Island would have little influence? The state also felt strongly that slavery should be abolished and a Bill of Rights be added to the Constitution. Consequently, the state has been unfairly characterized by many nineteenth-century historians as "Rogue Island" because of this defiant nature when it came to cooperating with the other twelve colonies in trying to establish a new Constitutional government.

These historians forget that Rhode Island was a strong supporter of the Revolutionary War and even declared its independence from Great Britain on May 4, 1776—two months before the Continental Congress declared its independence on July 4, 1776. Rhode Island's two delegates to that Continental Congress proudly signed their names to the declaration. Also, General Nathanael Greene, a native Rhode Islander, was one of George Washington's most trusted and competent generals, fighting many important battles in the South during the Revolutionary War.

On the night of June 9, 1772, Rhode Island Patriots successfully overtook the British ship *Gaspee*, which had been harassing both oceangoing American vessels as well as small boats ferrying goods from one side of Narragansett Bay to the other. Captain William Dudingston of the *Gaspee* had long suspected the Americans were taking smuggled goods off ships that had paid no duty tax and then moving the goods to area warehouses.

The American Patriots had successfully tricked the *Gaspee* to run aground on Namquit Point (now Gaspee Point) and, knowing that the tide would not come up to free the ship until early the next morning, took advantage of the opportunity to shoot the captain and overtake the ship from its sleepy crew before burning it to the water's edge. Dudingston was injured and carried to a house where his bloodstains on the floor were pointed out for a century thereafter as the "first blood of the Revolution."

The little state of Rhode Island, which measures just forty-eight miles long and thirty-seven miles wide, suffered and sacrificed much for the cause of American liberty. Its Patriots fought long and hard in many Revolutionary War battles and did all they could to thwart British wartime goals while occupying the state for three years during the Revolution.

BRISTOL

A gun house was built in 1808, but the location is unknown and it is no longer standing.

NEWPORT

Newport is a beautiful and historic old city founded in 1639 by Massachusetts refugees. It served as the state capital of Rhode Island for over 250 years before the seat of government was moved to Providence at the turn of the twentieth century. This city suffered extensively for three years during the occupation by British and Hessian troops at the start of the Revolutionary War. As a result, Newport lost much of its population, nearly all its commerce and many buildings. After the war, because Newport was heavily in debt, its people were too poor to tear down many of the existing old buildings in the name of progress. As a result, the city is now one of New England's most popular tourist destinations because of its large number of impressive Colonial, Federal and Victorian-era homes and mansions, among many other landmarks. Visitors to Newport should visit several historic sites, including Trinity Church, built in 1725–26, and noted for its tall, graceful spire and impressive triple-deck wineglass pulpit. Nor should they miss the Colony House, built in 1739–41, the second-oldest capital building in the United States. Here the Rhode Island legislature declared its independence from Great Britain on May 4, 1776. Here on the steps of the old state capitol is where the first reading of the Declaration of Independence in Rhode Island was delivered on July 20, 1776. Touro Synagogue, built in 1759, is the oldest house of Jewish worship in the United States and is noted for its simple yet elegant classical design. Presidents George Washington and Thomas Jefferson visited this synagogue, as did many other prominent men of the Revolutionary War era. It was in this synagogue that President Washington pledged religious freedom in the early days of the republic.

The Newport Powder House was built in 1755 on Gallows Field, designed to be eight feet square and twenty feet high. This powder house was used until February 1871, when a new one was built near Easton's Beach. No record could be found revealing when, where or how long this second powder house stood.

On December 3 (some records say December 5), 1774, the Rhode Island legislature ordered the removal of cannons and gunpowder from the powder house at Fort George. The orders were carried out on December 9, 1774. This was the first such move in America. Unlike the raid on and the capture of Fort William and Mary in New Castle, New Hampshire on December 14, 1774, the removal of powder and guns from Fort George was done without opposition from the British and was simply the confiscation of powder already in Patriots' possession. Therefore the Fort George powder acquisition does not constitute being "the first overt act of the American Revolution." The powder house no longer stands.

Another old Newport Powder House, date of construction unknown, stood on Rose Island and was later converted into a "smoke-less powder-dry house" circa 1902. No further information is available on this structure.

Fort Adams, built in 1799 and named in honor of President John Adams, stands on Ocean Drive in Newport. It was remodeled in 1824 and again circa 1945 and became the second largest bastioned fort in the United States. Designed to be the most heavily armed fort in America, it could garrison 2,400 troops at the time of construction. The fort consists of 21 acres, and its extensive defenses include powder magazines, breast-height walls, listening tunnels and three tiers of guns to defend the East Passage of Narragansett Bay.

PROVIDENCE

Providence was settled in 1636 by Roger Williams, who fled religious persecution from the strict Puritans in Massachusetts. His reasoning for naming the new settlement "Providence" is contained in a statement he wrote stating, "Having a sense of God's merciful providence unto me (I) call this place Providence. I desire it to be a shelter for persons distressed for (religious) conscience."

During the pre-Revolutionary period, England's policies were threatening the prosperity of this rapidly growing commercial city. The rum-slave-molasses trade, privateering and miscellaneous ship traffic to remote ports of call brought considerable wealth to Providence and turned it into a thriving seaport community of first rank. With much at stake when England's restrictive policies were threatening the prosperity of the American colonies and interfering with their commercial and political freedom, citizens of Providence engaged in several rebellious acts against the British Crown.

On March 2, 1775, Providence followed Boston's example and held a Tea Party, in which the whole town was invited to participate. A large pile of tea was heaped up in Market Square with a barrel of tar poured over it to ensure a good blazing fire, and then the top of the pile was crowned with a copy of Lord North's speech and the entire pyre ultimately lit, much to the delight of the local citizenry. On May 4, 1776, Rhode Island declared its independence from England when the Independence Act was signed in the old State House and read to townsmen from Jabez Bowen's balcony on

Market Square, where only sixteen years before, the accession to the British throne by George III had been proclaimed and celebrated. Rhode Island became the second state to declare its independence from England only after New Hampshire had done so on January 5, 1776.

During the Revolutionary War, Providence contributed money, ships, supplies and, by mid-1775, two thousand men to fight for the Patriot cause. No fighting took place in Providence during the war, but it was protected by a fort built near Robin Hill, and a lamp beacon was hung from the highest point of Prospect Hill to warn of any approaching British soldiers.

The Providence Powder House, built circa 1765, stood on Powder House Lane (now South Court Street) on the northwest corner of Prospect and Old Gaol Lane. It was repaired in 1781. According to an article in the Monday, September 2, 1765 *Providence Gazette*:

> *An aged Negro Woman was found dead in the fields at the new Powder House* [on Tuesday, August 27] *who had belonged to Mr. Isaac Depeyster (of Providence). The Coroner's Inquest found, that being notoriously addicted to Theft, Drunkenness, and other Vices, she had been long a Nuisance to her Master's Family: and having been out on Monday Evening, was refused Admittance, and died thro' old Age, and want of proper Nourishment.*

Isaac Depeyster was indicted on negligent homicide and found innocent, and so the old slave woman died as a result of exposure, starvation, neglect and old age.

Fort Independence was built on Robin's Hill near New York Avenue in Providence in 1775. Presumably, a powder house made up part of the adjoining complex of forts that protected the city up through the War of 1812. Today, only Fort Independence, which serves as a local park, survives.

Providence was also the site of Prospect Hill Fort, Fox Hill Fort, Hog Pen Point Fort and Fort Sullivan—all built and utilized during the Revolutionary War. Presumably, all possessed powder houses as part of their respective fortifications, but records fail to confirm their existence.

Tiverton

Tiverton—like its neighboring towns of Portsmouth, Middletown and Little Compton—suffered greatly from the oppressions of the British Army during the Revolutionary War.

Fort Barton, formerly located high on a Tiverton hill, was named after Colonel William Barton, who on the evening of July 9, 1777, along with forty Patriots, captured a stunned and undressed British general Richard Prescott in Portsmouth during one of the boldest military engagements executed during the entire Revolutionary War. Prescott begged to put clothes on but was denied and was hastily dragged through a field of thorny blackberry bushes. In August 1778, Generals John Sullivan, Nathanael Greene and Marquis de Lafayette came to this fort to view the impressive retreat of American Patriots during the Battle of Rhode Island, which was fought to the west on the hills of Portsmouth.

It is believed that Fort Barton may have had a powder house built during the Revolutionary War 1775–83, but it cannot be positively confirmed.

Warren

Warren sits on the east bank of the Warren River, a tributary of Narragansett Bay. It was incorporated as Sowams in 1668 and was a part of Massachusetts until 1747, when it was annexed to Rhode Island and renamed Warren. At a very early date, the town engaged in maritime pursuits and became a well-known whaling port. Shipbuilding also played a significant role in its early economy.

The outbreak of the Revolutionary War wreaked havoc on Warren's commercial prosperity. The town suffered much during the war and was raided by five hundred British and Hessian troops on May 25, 1778. They burned the Baptist meetinghouse, blew up the powder magazine (built prior to 1778), ransacked homes and businesses, ruthlessly destroyed property and sent terrified women and children screaming into the streets. A number of boats collected for use in an expedition against the British Army were burned and several local citizens taken as prisoners. From this terrible blow, Warren struggled to regain its former economic status. Within ten years after the close of the Revolution, commerce revived and Warren once again became an important shipbuilding community. Nothing else is known about Warren's powder house.

Rhode Island had several fortifications protecting various sections of Narragansett Bay dating back into the early 1700s. Many of these old forts presumably contained powder houses that are unfortunately not specifically documented.

CHAPTER 7
VERMONT POWDER HOUSES

Vermont entered the Union on March 4, 1791 as the fourteenth state. The Green Mountain state had a difficult beginning. The New Hampshire Grants, as Vermont was originally called, was fought over by both New Hampshire and New York, which claimed the disputed territory between the Connecticut River and twenty miles east of the Hudson River as part of their territorial domain. Starting in 1741, Governor Benning Wentworth of New Hampshire, a shrewd politician and businessman, charted 134 towns in the disputed territory and reserved for himself five hundred of the best acres in each town, getting very rich by doing so. These grants angered the governor of New York, who challenged New Hampshire's authority. In 1764, King George III stepped in and settled the boundary dispute in favor of New York, proclaiming that New Hampshire's territorial jurisdiction ended on the west bank of the Connecticut River. By then, many settlers and real estate speculators who had purchased New Hampshire titles to portions of Vermont became angry when they learned they had to pay New York to recognize their land deeds. Add this frustration to New York re-granting towns that were in direct conflict with earlier established towns by New Hampshire, and tempers began to flare. Enter the Green Mountain Boys, under the leadership of Ethan Allen, who began to physically expel New York sheriffs and surveyors from disputed towns.

After Allen captured Fort Ticonderoga in May 1775 for the American Army during the Revolutionary War, he hoped the Continental Congress would recognize the legitimacy of the New Hampshire Grants dispute. New York refused to cede its claims, and Congress declined to antagonize New York. Thus the stage was set for a group of plucky settlers to form a "constitutional convention" at a tavern in Windsor, officially declaring Vermont as an independent republic in July 1777. For fourteen years, Vermont governed itself by abolishing slavery, granting every man a vote (not just male landholders), establishing its own militia and printing its own money. The cocky little republic probably would have joined the Union in a heartbeat, but it was forced to chug along on its own for fourteen years. The new republic began to shift its boundaries by annexing parts of New Hampshire and New York and by threatening to rejoin the British empire by becoming a part of Canada. The Continental Congress became frustrated with Vermont and debated slicing Vermont down the middle and awarding the two pieces to New Hampshire and New York.

Finally, after many years of frustration and land-grabbing, a deal was cut in 1791 by the U.S. Congress in which Vermont paid a $30,000 monetary settlement to New York, which in turn withdrew its objections to Vermont becoming the fourteenth state. A new stripe was then added to the American flag, thus making fourteen stripes representing each state that comprised the Union at that time. The flag reverted back to just thirteen stripes in 1818, representing the original thirteen colonies.

Vermont played a significant role during the Revolutionary War, especially during the Battle of Bennington in August 1777, which actually took place just over the New York border. Mount Independence was another major Revolutionary War fort built on the Vermont side of Lake Champlain, where three thousand soldiers were stationed in early July 1777 in anticipation of a British attack under General John Burgoyne from the north. The attack never happened because the undermanned, poorly equipped fort was abandoned before the British could attack it. General John Burgoyne once described the people of Vermont as "the most rebellious race on the continent and [they] hang like a gathering storm upon my left." General Burgoyne, who was leading the 1777 British invasion from Canada to split the colonies along the Champlain and Hudson Valleys, was frustrated by the stubbornness of Vermont citizens, who refused to surrender to his invading army.

The only Revolutionary War battle to actually take place on Vermont soil happened on July 7, 1777, at Hubbardton when a successful rear-guard action by Colonel Seth Warner's Vermont, Massachusetts and New Hampshire troops ended a British pursuit under Generals Fraser and Reidesel. Though the Battle of Hubbardton was a loss for the Americans, it allowed General St. Clair's American army retreating from Mount (Fort) Independence and Fort Ticonderoga to escape and be able to fight once again near Bennington, Vermont, and Saratoga, New York. General Burgoyne's drive in 1777 to divide the colonies was first resisted at Hubbardton, which eventually ended in his defeat at the second Battle of

Saratoga in October 1777. The Battle of Hubbardton was considered one of the most successful rear guard battles in early American military history. Because the Americans fought so valiantly at Hubbardton, the well-equipped and better-trained British Army lost nearly 20 percent of its army to death and injuries.

Bennington

The historic city of Bennington dates back to 1749, when it became the first of many towns in present-day Vermont granted by New Hampshire provincial governor Benning Wentworth between 1749 and 1764.

The site of the two-story wooden Continental storehouse and/or possible ammunitions/powder house built prior to 1777 was close to where the impressive 1891 Bennington Battle Monument now stands. It was torn down in 1891 to make room for the battle monument and parking area. The wide-open vista surrounding part of the monument shows that this location was also very suitable for the grazing of cattle and the storage of food, which was greatly needed by British general John Burgoyne for his advancing army. The supplies in the Bennington Continental storehouse presumably included food, ammunition, saddles, clothes, blankets and other vital supplies typical for the use of an army on the move. These supplies quickly became the target of General Burgoyne, who wanted to capture them, leading to the Battle of Bennington on August 16, 1777. The battle resulted in a major victory for the Americans, thanks in part to New Hampshire general John Stark, who wisely suspected that Burgoyne was in dire need of additional supplies for his army and that the Continental storehouse at Bennington was going to be his main objective.

Castleton

Fort Warren was a fort located in Castleton from 1777 to 1779. American troops during the Revolutionary War retreated to Castleton after the recapture of Fort Ticonderoga by the British. Fort Warren was then built as part of a line of forts to defend Vermont. It is believed to have contained a powder house as part of the fort.

CORINTH

The Corinth Town History, published in 1964, contains an 1858 map of the Cookville section of Corinth that shows a powder house still in existence at that time. The town history reveals no further information on this powder house.

NEWBURY

The northern Vermont town of Newbury built a small brick powder house in 1809 among pine trees at the summit of Montebello, a small knoll located a short distance from Newbury Village. At a special town meeting in September 1810, it was voted to raise $150 to purchase ammunition to furnish the powder house. The powder house was struck by lightning when empty and rebuilt in 1836. This second powder house was still in existence at the outbreak of the Civil War in 1861 but had disappeared by 1902, when the town history was published.

Stereoview of Newbury, Vermont, circa 1860. This bucolic northern Vermont town formerly had a powder house dating back to 1809 that stood on Montebello Hill at least until 1861. *Matthew E. Thomas collection.*

ORWELL

Mount Independence has been called the most interesting and important historic site in Vermont and was the major Revolutionary War fortification built on Vermont soil. It has also been termed the least-disturbed major Revolutionary War site in the United States today.

The fort (called Mount Independence) sits atop a rocky promontory overlooking Lake Champlain directly opposite Fort Ticonderoga (located on the New York side of Lake Champlain) and is protected on three sides by the lake and by steep cliffs towering two hundred feet in elevation. During construction of the fort, news of the signing of the Declaration of Independence reached the fort on July 18, 1776, and following the reading of the Declaration of Independence on July 28, 1776, the troops jubilantly christened the new fortification "Mount Independence."

Expecting a British attack, Congress had authorized the fort to be built to maximum strength so that troops from several colonies totaling over twelve thousand men in four brigades—three of which were stationed at "the Mount"—could be accommodated within the fort. The foundation of a building, either a powder magazine or quarters for a gun crew, is located behind the southern battery within the fort. It is one of the best preserved foundations on Mount Independence. The formidable appearance of the two forts with flags flying and artillery prominently displayed, along with the approaching winter, persuaded a British commander to withdraw his powerful forces to Canada for the season. Mount Independence and Fort Ticonderoga had accomplished their mission of defending the northern frontier with scarcely a shot fired. With the departure of the enemy, the population of the fort was reduced to a minimum for the winter. However, the suffering that had accompanied the building of the fort, such as fever and smallpox, was augmented by the cold and the lack of supplies. One night, several men froze to death in their huts, and they were but a few of the hundreds who died at Mount Independence from exposure and disease.

Despite these conditions, work was started in February 1777 on a great floating bridge connecting Fort Ticonderoga and Mount Independence by building log caissons on the ice, filling them with rock and then sinking them through the ice in the spring for use as bridge anchors. Work continued with the arrival of warmer weather; the hospital was built in April 1777, and the great floating bridge and boom were secured about the same time.

In June 1777, reports were again received of a major British advance from the north, this time with a force of about seven thousand men under the command of General John Burgoyne. American leaders had expected a campaign to be launched from British-held New York City and had not provided adequate manning in the north at Mount Independence or Fort Ticonderoga. Instead of the necessary ten thousand troops required to man the forts properly, only about three thousand poorly equipped men remained to face the oncoming British attack. American general Arthur St. Clair—faced with an undermanned and undersupplied fort complex, imminent bombardment and the approaching cutoff of his retreat—ordered the immediate withdrawal from both forts during the night of July 5–6, 1777. The careless burning of a building on Mount Independence alerted the British to the American withdrawal, and the redcoats then pursued the Americans to Hubbardton, Vermont, where the British were checked in a bloody battle by the American rear-guard, saving the main American force to face General Burgoyne in the American's decisive victory at Saratoga, New York, a few months later.

Except for a short, unsuccessful American counter attack in September 1777, Mount Independence saw little further action. The surrender of Burgoyne at Saratoga on October 17, 1777, left the British garrison occupying Mount Independence and Fort Ticonderoga in an untenable position, and on November 8, 1777, it evacuated after burning the Mount Independence buildings and fortifications. The destruction of this Revolutionary War landmark marked the final chapter of one of Vermont's most significant contributions toward advancing the Patriot cause of American Independence.

Mount Independence is one of the most significant and fascinating Revolutionary War landmarks remaining in New England. Unfortunately, due to its relative historical obscurity, it is also one of the least known and appreciated sites associated with the American Revolutionary War. Visitors to Mount Independence—a National Historic Landmark—will most likely be pleasantly surprised and enthralled while touring one of the best-preserved Revolutionary War archaeological sites in America, consisting of approximately four hundred acres.

Pittsford

Fort Vengeance received its name after a soldier was killed by Indians shortly after the fort was first occupied in 1780. His comrades swore vengeance on the Indians and thus the origin of the fort's name. The powder house believed to have been a part of Fort Vengeance no longer stands.

RUTLAND

A powder house believed built as part of Fort Rutland in 1775 no longer stands. Another powder house, built as part of Fort Ranger when the fort was built in 1778 in a section of Rutland called Mead's Falls, also no longer stands.

SHAFTSBURY

Powder and ammunition was stored in a somewhat fortress-like structure built in 1770 in the rural countryside of Shaftsbury in preparation for the Battle of Bennington, Vermont, which took place on August 16, 1777. The "Munitions House," as it is called, stands as a private residence about one hundred yards east of Route 7A on Buck Hill Road in South Shaftsbury.

On a side note, the famous Bennington Battle Flag—designed with its alternating red and white stripes representing the thirteen colonies and the numeral "76" set in a blue canton with thirteen seven-pointed stars—has

This half stone, half wood 1770 powder house/ammunitions magazine at Shaftsbury, Vermont, was used during the 1777 Battle of Bennington, Vermont. *Courtesy of the Vermont Historical Society.*

The 1770 powder house at Shaftsbury as it appears in 2010 after renovations as a private residence. *Author's photo.*

not been proven to have been the flag carried by the Green Mountain Boys and others against the British Army at the Revolutionary War Battle of Bennington in 1777. There is no historical reference to a flag with the stars-and-stripes design being carried in the Battle of Bennington. In actuality, there is no record of the Bennington flag being in existence prior to the late nineteenth century, when it was displayed at the Chicago Public Library. That flag shows it to have been made of cotton material and cotton thread, and the use of the numeral "76" is more typical of flags produced in the 1820s. It was also believed to have been made on a power loom, unknown in America before 1813.

It is therefore believed by many historians that the Bennington flag may have been actually designed for an 1824 visit to the United States by General Lafayette or for the fiftieth anniversary of the U.S. Declaration of Independence in 1826. The only documented flag used by American military forces at the Battle of Bennington was green with thirteen white five-pointed stars set in a blue canton carried by the Green Mountain Boys and others.

There were several forts scattered around Vermont. Most likely there were other powder houses with some of them, but records fail to reveal which ones.

BIBLIOGRAPHY

Authelet, Jack. *Glimpses of Early Foxboro, MA.* Foxboro, MA: Foxboro Company, 1978.

Baker, Edward. "Benedict Arnold Turns & Burns New London, CT." *Hog River Journal*, Fall 2006.

Barber, John Warner. *Connecticut Historical Collections.* New Haven, CT: Durrie & Peck & J.W. Barber, 1836.

Barber, Samuel. *Boston Common–A Diary of Notable Events, Incidents, and Neighboring Occurrences.* Boston: Christopher Publishing House, 1916.

Bearse, Ray. *Vermont: A Guide to the Green Mountain State.* Boston: Houghton Mifflin Company, 1966.

Bell, Charles H. *History of Exeter, NH.* Boston: Press of J.E. Farwell & Co., 1888.

Benedetto, Christopher. "An Awful and Tragic Scene." *New England Ancestors Magazine*, Summer 2007

Benedict, Reverend William A., and Reverend Hiram A. Tracy. *History of Sutton, MA, 1704–1876.* Worcester, MA: Sanford & Co., 1878.

Bittinger, Reverend J.Q. *History of Haverhill, N.H.* Haverhill, NH: Cohos Steam Press, 1888.

Blake, Jonathan. *History of the Town of Warwick, MA.* Boston: Noyes, Holmes & Co., 1873.

Boston (MA) Columbian Sentinel. "Distressing Event." July 8, 1809.

Bouve, Edward T. *History of Hingham, MA.* Vol. I. Published by the Town of Hingham. Cambridge, MA: University Press, 1893.

Brown, Warren. *History of Hampton Falls, N.H.* Manchester, NH: John B. Clarke Co., 1900.

Caulkins, Frances M. *History of New London, CT.* Hartford, CT: Press of Tiffany Case, 1860.

Chaffin, William L. *History of Easton, MA.* Cambridge, MA: University Press, 1886.

Chandler, Philemon. Diary. Manuscript in Andover, Massachusetts Historical Society.

Chase, George Winthrop. *History of Haverhill, MA.* self-published, 1861.

Child, Frank Samuel. *An Old New England Town (Fairfield, CT).* New York: Charles Scribner's Sons, 1895.

Childs, Ethel B. *History of Stow, MA. 1683–1983.* Stow, MA: Stow Historical Society Publishing Co., 1983

Clark, Charles E., James S. Leamon and Karen Bowden. *Maine: In The Early Republic.* London: University Press of New England, 1988. Published for the Maine Historical Society and Maine Humanities Council.

Clarke, George K. *History of Needham, MA.* Cambridge, MA: University Press, 1912.

Clark, George F. *History of Norton, MA.* Boston: Crosby, Nichols & Co., 1859.

Coburn, Silas Roger. *History of Dracut, MA.* Lowell, MA: Courier-Citizen Press, 1922.

Coffin, Joshua. *History of Newbury, Newburyport & West Newbury (MA.)* Hampton, NH: Peter E. Randall, Publisher, 1977.

Connecticut Historical Commission. *Connecticut History & Culture, 350 Years, 1635–1985.* Willimantic, CT: Center of Connecticut Studies, Eastern Connecticut State University, 1986.

Coolidge, Mabel Cook. *History of Petersham, MA 1637–1947.* Hudson, MA: Powell Press, 1948.

Crofut, Florence S. Marcy. *Guide to the History & Historic Cities of Connecticut.* Vols. I & II. New Haven, CT: Yale University Press, 1937.

Currier, John J. *History of Newbury, MA 1635–1902.* Boston: Damrell & Upham Publishers, 1902.

———. *Newburyport, MA. 1764–1905.* Newburyport, MA: Damrell & Upham Publishers, 1906.

Cutter, Benjamin, and William R. Cutter. *History of Arlington, MA.* Boston: David Clapp & Son, 1880.

Daggett, John. *History of Attleborough, MA.* Boston: S. Usher Publisher, 1894.

Daniels, George F. *History of Oxford, MA.* Worcester, MA: Charles Hamilton Press, 1892.

DeForest, H.P. *History of Westborough, MA.* Cambridge, MA: University Press, 1891.

DeLue, Willard. *Story of Walpole, MA.* Norwood, MA: Ambrose Press, 1925.

Dobbs, Judy D. *A History of the Watertown Arsenal, 1816–1967.* Watertown, MA: U.S. Army Materials and Mechanics Research Center, 1977.

Drake, Samuel Adams. *Fourth Report Boston Records: Town of Roxbury, MA*. Boston: Municipal Printing Office, 1905.

———. *History of Middlesex County, MA*. Boston: 1880.

———. *Old Landmarks and Historic Personages of Boston.* Rutland, VT: Charles E. Tuttle Co., 1872.

Drumm, Nelde K., and Margaret P. Harley. *The Heritage of Turkey Hills 1718–1978.* Lunenburg, MA Historical Society. Leominster, MA: A&L Graphics, 1977.

Earls, Alan R. Images of America: History of the Watertown Arsenal. Charleston, SC: Arcadia Publishing, 2007.

Emery, Helen F. *A Puritan Village Evolves: A History of Wayland, MA*. Canaan, NH: Phoenix Publishing, 1981.

Emery, Samuel H. *History of Taunton, MA*. Syracuse, NY: D. Mason & Co. Publishers, 1893.

Encyclopedia of the United States. *Encyclopedia of Massachusetts.* New York: Somerset Publishers, Inc., 1984.

Felt, Joseph B. *Annals of Salem, MA*. Boston, MA: W.S.B. Ives Publishers, 1849.

———. *History of Ipswich, Essex & Hamilton, MA*. Ipswich, MA: Clamshell Press, 1966.

Fennelly, Catherine. *Old Sturbridge Village: A Guidebook.* Meriden CT: Gravure Publishing Company, 1966.

Frizzell, Martha. *History of Charlestown, NH.* Littleton, NH: Courier Printing Co., no date.

Fuess, Claude M. *Andover (MA.) Symbol of New England:The Evolution of a Town.* Portland, ME: Anthoensen Press, 1959.

Gallerani, Kathryn. "Wicked Local Kingston: Powder House Moved." *Kingston (RI) Reporter*, February 14, 2012.

Gillespie, C. Bancroft, and George Munson Curtis. *A Century of Meriden, CT.* Meriden, CT: Journal Publishing Co., 1906.

Groff, Bethany. *Brief History of Old Newbury, MA: From Settlement to Separation.* Charleston, SC: The History Press, 2008.

Gurney, Caleb S. *Portsmouth, NH: Historic & Picturesque.* Hampton, NH: Peter E. Randall, Publisher, 1981.

Harpin, Mathias P. *Harpin's Connecticut Almanac, Windham County Edition, 1970–1971*. Oneco, CT: Harpin's Publishing Co., n.d.

Harris, Edward M. *Andover MA in the American Revolution.* Marceline, MO: Walsworth Publishing CO., 1976.

Harris, John. *America Rebels.* A Boston Globe Bicentennial Publication. Affiliated Publications, Inc., 1976.

Herring, Stephen W. *South Middlesex: A New England Heritage.* Northridge, CA: Windsor Publications, Inc., 1986.

Heywood, William S. *History of Weston, MA. 1728–1893.* Lowell, MA: Vox Populi Press, 1893.

Hinman, R.R. *Catalogue of the First Puritan Settlers of Connecticut.* Hartford, CT: Tiffany Case Publisher, 1852.

Historical Guide of Rehoboth, MA. May 1995.

"History of the Fort at Mount Independence State Historic Park, Orwell, VT." State of Vermont Tourism Bureau Booklet, 2010.

Holliston Historical Society. *Images of America: Holliston.* Dover, NH: Arcadia Publishing Dover, 1997.

Houghton, Raymond C. *A Revolutionary War Road Trip on U.S. Route 7 (VT).* Delmar, NY: Cyber Haus Publishers, 2004.

Howell, John Mead. *Lost Examples of Colonial Architecture Plate 240.* New York: W. Helburn Publishers, Inc., 1931.

Hudson, Alfred Sereno. *History of Sudbury, MA 1638-1889.* Sudbury, MA: Sudbury Press, 1968.

Hulbert, Joanne. *Holliston, MA: A Good Town.* Stonington, ME: Penobscot Press, 2000.

Huntoon, Daniel T.V. *History of Canton, MA.* Cambridge, MA: University Press, 1893.

Jackson, Francis. *History of Newton, MA. 1639–1800.* Boston, MA: Stacy & Richardson Printers, 1854.

Jewett, Amos Everett. *History of Rowley, MA. 1639–1850.* Rowley, MA: Jewett Family of America, 1946.

Karlson, Douglas. "The Redcoats Are Coming." *Harwich (MA) Oracle*, August 16, 2005.

Lawry, Nelson H., Glen M. Williford and Leo K. Polaski. *Images of America: Portsmouth (N.H.) Harbor's Military & Naval Heritage.* Charleston, SC: Arcadia Publishing, 2004.

Lewis, A., and J.R. Newhall. *History of Lynn, MA.* Boston: Shorey Publisher, 1865.

Lincoln, William, and C.C. Balwin. *Worcester Magazine & Historical Journal.* Worcester, MA: Charles Griffin Printers, 1826.

Locke, Emma P. *Colonial Amherst, N.H.* Milford, NH: W.B. & A.B. Rotch Press, 1916.

Lord, Priscilla S., and Virginia C. Gamage. *History of Marblehead, MA.* New York: Chilton Book Co., 1972.

Love, Reverend William DeLoss. *Colonial History of Hartford, CT.* U.S. Bicentennial Series. State of Connecticut, 1974.

Lowell Sun. "Powder Magazine Explosion on July 29, & 30, 1903." July 30, 1903.

Lynnfield, Massachusetts Heritage Association. *Lynnfield, MA: 1895–1976.* Canaan, NH: Phoenix Publishing Co., 1977.

Manchester (NH) Union Leader. "Powder Blast Rocks City." July 4, 1944.

Mason, Orion T. *Handbook of Medway, MA. History 1713–1913.*

Massachusetts Bicentennial Commission. *Commemorative Guide (to the) Massachusetts Bicentennial.* Dublin, NH: Yankee, Inc., 1975.

McDuffee, Franklin. *History of Rochester, N.H. 1722–1890.* 2 vols. Manchester, NH: John B. Clarke, 1892.

McGarry, Carol. *History of the Whitney-Paul Tavern, Sherborn, MA.* Dover, MA: Dover-Sherborn Press, 2001.

Merrill, Joseph. *History of Amesbury and Merrimac, Massachusetts.* Haverhill, MA: Press of Franklin P. Stiles, 1880.

Merrill, Nancy. "Exeter, N.H. Powder House." *Exter Newsletter*, November 20, 1974.

Moore, J. Bailey. *History of Candia, NH.* Manchester, NH: George W. Browne, 1893.

Murphy, Rosalind. *Powder House on Prospect Hill in Beverly, MA.* Privately published, circa 2012.

Murray, Barbara, Deborah E. Johnson and Jayme Kulesz. *Groton (MA) at 350.* Published by the Town of Groton, MA, 2005.

Nashua (NH) Weekly Telegraph. "Portsmouth, N.H. Powder House," June 3, 1876.

Nelson, Charles B. *History of Stratham, N.H. 1631–1900.* Somersworth, NH: New Hampshire Publishing Co., 1965.

New York Times. "Bridgeport/West Stratford, CT Powder House Explosion," April 22, 1881.

Old Records of Fitchburg, MA, Volume IV. 1796–1809. Published by the Fitchburg, MA Historical Society 1901.

Osterweis, Rollin G. *Three Centuries of New Haven,* Connecticut *1638–1938: The Tercentennary History.* New Haven, CT: Yale University Press, 1953.

Page, Elwin L. "The King's Powder, 1774." *New England Quarterly*, n.d.

Paige, Lucius. *History of Cambridge, MA, 1630–1877.* Cambridge, MA: 1877.

Parsons, Charles L. *The Capture of Fort William & Mary.* Reprinted from Proceedings of the N.H. Historical Society, 1974.

Parsons, Langdon B. *History of Rye, N.H. 1623–1903.* Concord, NH: Rumford Printing Co., 1905.

Pattee, W.S. *History of Old Braintree & Quincy, MA.* Quincy, MA: 1878.

Plainfield, New Hampshire Historical Society. *Choice White Pines & Good Land: A History of Plainfield & Meridian, N.H.* Portsmouth, NH: Peter Randall Publisher, 1991.

Providence (RI) Gazette. "Providence Powder House." September 12, 1765.

Purington, Dott. *A Candia (N.H.) Collection.* Privately published, 2005.

Records of Duxbury, MA, 1642–1770. Published by Town of Duxbury. Plymouth, MA: Avery & Doten, 1893.

Records of Town of Plymouth, MA. Vol. III, 1743–83. Plymouth, MA: Memorial Press, 1903.

Register of Soldiers, Sailors and Patriots. Sons of American Revolution Massachusetts Society, 1904.

Report of the Chief of the Bureau of Ordnance to the Secretary of the Navy. 1906.

Residents and Business Directory of Foxborough, MA, 1887–88.

Ripley, Emma F. *Weston, MA: A Puritan Town*. Cambridge, MA: Riverside Press, 1961.

Rossiter, William S. *Days & Ways In Old Boston*. Boston: R.H. Stearns & Co., 1915.

Roth, David M. *Connecticut: A History*. New York: W.W. Norton & Company, 1979.

Runnels, Moses Thurston. *History of Sanbornton, N.H.* 2 vols. Boston, MA: Alfred Mudge & Son, 1881–82.

Samuels, Edward A., and H.H. Kimball. *MA: Past & Present*. Boston, MA: self-published, 1897.

Scales, John. *History of Dover, NH*. Manchester, NH: John B. Clarke Co., 1923.

Schenck, E.H. *History of Fairfield, CT*. Astor Place, NY: J.J. Little & Co., 1904.

Schmidt, Jay. *Fort Warren: New England's Most Historic Civil War Site*. Amherst, NH: UBT Press, 2003.

Scribner Historic Buildings Series. *Historic Buildings of Massachusetts—Photographs from the Historic American Buildings Survey*. New York: Charles Scribner's Sons, 1976.

Secomb, Daniel F. *History of the Town of Amherst, NH*. Concord, NH: Evans, Sleeper & Woodbury Press, 1883.

Sewall, Samuel. *History of Woburn, MA*. Boston: Wiggin & Lunt Publishers, 1868.

Shaughnessy, Anne Carr. *A Guide To Sherborn, MA. 300th Anniversary Committee 1974*. A digital version prepared by the Sherborn, Massachusetts Public Library in 2007.

Smith, Frank. *History of Dedham, MA*. Dedham, MA: 1936.

Smith, J.E.A. *History of Pittsfield, MA*. Springfield, MA: C.W. Bryan & Co., 1876.

Smith, Joshua M. *Blockhouse & Battery: A History of Fort Edgecomb*. Published by Friends of Fort Edgecomb, 2009.

Smith, Richard B. *The Revolutionary War in Bennington County, VT*. Charleston, SC: The History Press, 2008.

Smith, S.F. *History of Newton, MA*. Boston: American Logotype Co., 1880.

Smith, Tyler. "Fort Sewall Has Many Stories to Tell." *Marblehead (MA) Reporter*, November 17, 2008.

Spiller, Virginia S. *Historic York, Maine Colonial Ramblings*. Privately published, 1998.

Stahl, Jasper J. *History of Old Broad Bay & Waldoboro, ME*. Portland, ME: Bond Wheelwright Company, 1956.

Stanley, William "Bill." "Norwichtown to Come Alive with History." *Norwich Bulletin*, July 6, 2008.

Starkey, Marion L. *A Little Rebellion*. New York: Alfred A. Knoph, 1955.

Stearns, Ezra. *History of Ashburnham, MA, 1734–1886*. Town of Ashburnham, MA, 1887.

Tappan, William H. *History of Manchester, MA.* Philadelphia: J.W. Lewis & Co., 1888.

Temple, J.H. *History of North Brookfield, MA.* Boston: Rand Avery Co. Publisher, 1887.

Thomas, Matthew E. *History of Fremont, NH—Olde Poplin—An Independent New England Republic 1764–2004.* Manchester, NH: New England Duplicator & Printing, 2004.

———. *Images of America: Rockingham County, New Hampshire.* Augusta, ME: Alan Sutton Publishers Inc., 1994.

———. *Original Territory of Historic Exeter, N.H.* Portsmouth, NH: Randall Press, 1976.

300th Anniversary Booklet of Sherborn, MA, 1974. Digital version prepared by Sherborn Public Library, Sherborn, Massachusetts, 2007. Tilden, William S. *History of Medfield, MA.* Boston: George H. Ellis Publisher, 1887.

Tree, Christina, and Elizabeth Roundy. *Maine: An Explorers Guide.* Woodstock, VT: Countryman Press, 1995.

Trumbull, James Hammond. *Memorial History of Hartford County, Connecticut 1633–1884.* Boston: E.L. Osgood Publisher, 1886.

250th Anniversary of Haverhill, MA. Boston: Joseph George Cupples Printer, 1891.

Van Dusen, Albert E. *Connecticut: A Fully Illustrated History of the State from the Seventeenth Century to the Present.* New York: Random House, 1961.

Waldo, George Curtis. *History of Bridgeport, CT.* Hartford, CT: S.J. Clarke, Publisher, 1917.

Wallace, James Burns. *History of Canaan, N.H.* Concord, NH: Rumford Press, 1910.

Warren, Bill. *Rye, NH Reflections: A Visual History.* Published privately, 2007.

Waters, Thomas F. *Ipswich in the Massachusetts Bay Colony.* Vols. I & II. Ipswich, MA: published privately, 1917.

Watkins, Lura Woodside. *Middleton, Massachusetts: A Cultural History.* Boston: Henry N. Sawyer Co., 1970. Published for Essex Institute, Salem, MA.

Wells, Frederic P. *History of Newbury, VT.* St. Johnsbury, VT: Caladonian Co., 1902.

Westport, MA Historical Society. Westport, MA Powder House Dedication Program, June 3, 1972.

Wheeler, Ruth R. *Concord, MA: Climate for Freedom.* Concord Antiquarian Society, Yarmouthport, MA Printing & Publishing Associates, 1967.

Whitehill, Walter M., and Norma Kotker. *Massachusetts: A Pictorial History.* New York: Charles Scribner & Sons, 1976.

Whitehill, Walter Muir. *Boston: A Topographical History.* Cambridge, MA: Belknap Press of Harvard University Press, 1963.

WPA Federal Writer's Project. *Connecticut: American Guide Series.* Cambridge, MA: Houghton Mifflin Co., 1938.

———. *Maine: American Guide Series.* Cambridge, MA: Houghton Mifflin Co., 1937.

———. *Massachusetts: American Guide Series.* Cambridge, MA: Houghton Mifflin Co., 1937.

———. *New Hampshire: American Guide Series.* Cambridge, MA: Houghton Mifflin Co., 1938.

———. *Rhode Island: American Guide Series.* Cambridge, MA: Houghton Mifflin Co., 1937.

———. *Vermont: American Guide Series.* Cambridge, MA: Houghton Mifflin Co., 1937.

Wikipedia was consulted while researching some of these old powder houses, powder magazines and gun houses.

INDEX

C

D

E

F

G

H

I

K

L

M

N

O

P

Q

R

ABOUT THE AUTHOR

Matthew E. Thomas developed a passion for American history as a young adolescent and has researched, written and published a dozen books and pamphlets on early New England history. He studied architecture and archaeology at the College of William & Mary in Williamsburg, Virginia.

Over the past four decades, he has devoted his life to researching, writing, promoting and preserving New England's colorful heritage. He founded New England Historical Research Associates in 1978, where he provided lectures, seminars, historical and genealogical research and consulting services for over twenty-five years. He served as past secretary of the Association of New Hampshire Historical Societies and has been a member of the Organization of American Historians and the New Hampshire Historical Society. Mr. Thomas currently serves as president and curator of the Fremont, New Hampshire Historical Society. Between 1988 and 1992, he was instrumental in helping to restore and preserve the American Independence Museum in Exeter, New Hampshire, and the unique 1800 Twin-porch Meeting House in Fremont.

Mr. Thomas serves as a respected historical consultant to numerous authors, historians, educators, genealogists, architects, archaeologists, filmmakers, historical re-enactors, museums, schools, universities and colleges nationwide.

He and his wife, Sharon, have played an active role in a variety of capacities by helping to preserve the history of their beloved hometown of Fremont, New Hampshire.

They are particularly blessed to be the proud parents of a daughter, Rebecca, grateful in-laws to Daniel and loving grandparents to Callie.